Getting Inside Your Head

What Cognitive Science
Can Tell Us about Popular Culture

LISA ZUNSHINE

The Johns Hopkins University Press
Baltimore

KH

The Johns Hopkins University Press
2715 North Charles Street
Baltimore, Maryland 21218-4363
www.press.jhu.edu

Library of Congress Cataloging-in-Publication Data

Zunshine, Lisa.
  Getting inside your head : what cognitive science can tell us about popular culture /
Lisa Zunshine.
    p. cm.
  Includes bibliographical references and index.
  ISBN 978-1-4214-0616-9 (hdbk. : acid-free paper) — ISBN 1-4214-0616-0 (hdbk. : acid-free
paper)
  1. Psychology and literature. 2. Cognition and culture. 3. Popular culture and literature.
4. Characters and characteristics in literature. 5. Philosophy and cognitive science. I. Title.
  PN56.P93Z86 2012
  801'.92—dc23   2011048290

A catalog record for this book is available from the British Library.

*Special discounts are available for bulk purchases of this book. For more information, please
contact Special Sales at 410-516-6936 or specialsales@press.jhu.edu.*

The Johns Hopkins University Press uses environmentally friendly book materials,
including recycled text paper that is composed of at least 30 percent post-consumer waste,
whenever possible.

6/25/13

But what was to be done about the impossibility of seeing into other people's souls?
—Ellen Spolsky, *Word vs Image*

Contents

||||||||||||||||||||||||||||||||||||||||||||||||||||||||||||||||||||||||||||||||||||||||||||||||||||||||||||||||||||||||

Illustrations

Preface: Fantasies of Access

||||||||||||||||||||||||||||||||||||||||||||||||||||||||||||||||||||||||||||||||||||||||||||||||||||||||||||||||||||||||||||||

In which Bart Simpson is thinking; Mona Lisa is smiling; the purpose of this book is revealed; its genre is canvassed; and a long-due gratitude is expressed.

We live in other people's heads: avidly, reluctantly, consciously, unawares, mistakenly, inescapably. Our social life is a constant negotiation among what we think we know about each other's thoughts and feelings, what we want each other to think we know, and what we would dearly love to know but don't.

We've been doing this for hundreds of thousands of years. Cognitive scientists have a special term for the evolved cognitive adaptation that makes us attribute mental states to ourselves and to other people; they call it theory of mind or mind reading. Though it may sound like telepathy, theory of mind is actually its opposite. Telepathy implies perfect self-conscious access to someone's thinking. Mind reading is approximate guessing and imperfect interpretation, most of it taking place below the radar of our consciousness.

Our culture is both a product of theory of mind and its stomping ground. We enter it by attributing mental states to everybody from Bart Simpson to Plato and from Mona Lisa to the drafters of the U.S. Constitution. Cultural representations, high and low, exploit the fact that we live in other people's heads yet have no direct access to their thoughts and feelings. Novels, movies, paintings, and situation comedies all build on theory of mind, experiment with it, and feed it elaborate social fantasies.

One such fantasy is the fantasy of perfect access to mind through body. We get to see fictional characters at the exact moment when their body language betrays their real feelings. This is in contrast to real life, in which there is always a possibility that we will misinterpret seemingly transparent body language, particularly in a complex social situation, or that people will perform transparent body language to influence our perception of their mental states.

The fantasy of perfect access to mind through body is old, but it takes surprising new forms in different historical periods and genres: thirteenth-century Chinese operas; medieval ribald tales; eighteenth-century French paintings; nineteenth-century English novels; twentieth-century movies, musicals, photography, and stand-up comedy; and twenty-first century reality television. This book is about how this fantasy works, when it stops working, and why we can't get enough of it.

Though dealing with novels, film, and art, this is neither literary or film criticism nor art history. I don't offer a comprehensive analysis of a particular writer, movie, painting, genre, or motif, and I steer clear of specialized vocabulary. Most works under discussion have been extensively analyzed by others. Whenever I can, I happily point to instances of compatibility with existing studies, but, fruitful as I believe it to be, a sustained exploration of such compatibility is beyond the scope of this project. Pressed for the genre of what I do here, I would call it "cognitive cultural studies," though I think of it mainly as a book-length thought experiment: an attempt to view a variety of cultural phenomena from one particular perspective made possible by research in cognitive science and to push that perspective as far as possible.[1]

Writing up this thought experiment took five years and as many drafts, and I have incurred many intellectual debts along the way. As always, I found support and inspiration among the community of scholars working with cognitive approaches to literature and culture: Porter Abbott, Frederick Luis Aldama, Mike S. Austin, Elaine Auyoung, Joseph Bizup, Mary Crane, Nancy Easterlin, William Flesch, Monika Fludernik, F. Elizabeth Hart, David Herman, Patrick Colm Hogan, Tony Jackson, Suzanne Keen, Jonathan Kramnick, Howard Mancing, Bruce McConachie, Alan Palmer, Isabel Jaén Portillo, Alan Richardson, Elaine Scarry, Vernon Shetley, Ellen Spolsky, Gabrielle Starr, Simon Stern, and Blakey Vermeule.

I am grateful to the Guggenheim Foundation and to the University of Kentucky College of Arts and Sciences, whose generous support enabled me to spend a year and a half as a visiting scholar at the Yale Department of Psychology; to Paul Bloom, who made that magical year possible by inviting me to his Mind and Development Lab and who helped me realize the importance of restraint for embodied transparency; to the affiliates of his lab, particularly Mark Sheskin, for his valuable insights about Stephen Sondheim; to the members of the Mind, Brain, Culture, and Consciousness group at the Whitney Humanities Center at Yale; to Michael Holquist, Doug Whalen, Philip Rubin, Ken Pugh, and other members of the Teagle-Haskins Collegium; to Kang-i Sun Chang, for her generous help with the Chinese opera; to Stephen Kern for his similarly generous assistance with proposal compositions; to Evelyn Birge Vitz, whose inspiring approach to performance turned me to embodied transparency in medieval literature; to James Phelan, for his crucial early advice about transparent bodies and narrative; to Ralph James Savarese for introducing me to disability studies sensitive to the possibilities opened by the autistic view of the world, as opposed to just its limitations, and for stepping in at the last moment to correct relevant parts of my argument; and to Jason Flahardy at the University of Kentucky Special Collections and Kathryn Wong Rutledge and Mary Lou Cahal from the University of Kentucky Teaching Academic Support Center for their invaluable work on illustrations.

At the Johns Hopkins University Press I am grateful to Trevor Lipscombe, Matt McAdam, and Deborah Bors. I also thank my copyeditor Joe Abbott, and, most important, the anonymous reader whose comments made me revise large portions of the book.

Finally, I thank Etel Sverdlov for regularly offering words of wisdom and comfort, and Joel Kniaz for his incredible patience in offering detailed conceptual feedback on several different versions of this manuscript.

Getting Inside Your Head

ONE

{
In which the author first tries to read the mind of a stranger at the library
and then realizes that she doesn't know how to talk about it / an Israeli
immigration clerk makes a surprising gesture with her hand / emperor
Caracalla feels threatened / a British soccer player hopes to be mobbed
by his teammates / a tightrope walker faces competition from a Whee-lo
toy / Mona Lisa keeps smiling / and Andy Kaufman looks sincere.
}

Culture of Greedy
Mind Readers

I am writing this in a quiet library hall lined with long desks. In front of me I see a young woman turning around and glancing at the three whispering and occasionally laughing students to her left. I think the noise bothers her, and she wants to show it. But I could be wrong. Perhaps, bored after hours of sitting still, she appreciates this momentary diversion and wants to see its source. Or perhaps she wonders if she knows any of them. Or perhaps she is a sociologist and something about their group dynamics has caught her attention. I don't know her, so I am not likely ever to find out what she is actually thinking as she turns around. Still, I automatically interpret her body language in terms of her unobservable thoughts and feelings: she feels this; she wants that; she wants them to think that she thinks this or that; she wants other people to know that she is responding to that group's behavior.

And you are not in the least surprised by my reasoning. You, too, take it for granted that there must be some thought, desire, or intention behind her body language. Our everyday social interactions are unimaginable without this kind of intuitive reasoning: to make sense of any human action, we must see it in terms of a mental state that prompted it.

We've been doing this day and night for hundreds of thousands of years. (At night we attribute intentions to creatures populating our dreams.) Psychologists have a special term for the evolved cognitive adaptation that makes us see behavior as caused by underlying mental states. They call it *theory of mind,* also known as folk psychology and mind reading. The latter term is particularly inapt. Given how many of our attributions and interpretations of thoughts and feelings are wrong or only approximately correct, they might as well call it mind misreading. But since evolution doesn't deal in perfection, we have to fumble through by "reading minds" as best we can.

In the last five years theory of mind has become a major research topic among cognitive, developmental, comparative, and social psychologists, as well as cognitive neuroscientists. Though everything they learn opens up more questions and will remain the subject of debates for years to come, theory of mind is increasingly thought of as a crucial cognitive endowment of our species—a cornerstone of imagination, pretense, morality, and language, indeed of every aspect of human sociality.

As a cognitive adaptation, mind-reading ability may have developed during the Pleistocene period, from 1.8 million to 10 thousand years ago. According to evolutionary psychologist Simon Baron-Cohen, the emergence of theory of mind was evolution's answer to the "staggeringly complex" challenge faced by our ancestors, who needed to make sense of the behavior of other people in their group, which could include up to two hundred individuals. As Baron-Cohen points out, "Attributing mental states to a complex system (such as a human being) is by far the easiest way of understanding it," that is, of "coming up with an explanation of the complex system's behavior and predicting what it will do next."[1]

Studies in theory of mind suggest a new way of understanding what constitutes our human environment. Usually, the word *environment* brings to mind trees, air, water, roads, houses, and such. If we remember, however, that the human species is foremost a *social* species—that is, our need and ability to communicate with others underlies every aspect of our existence—we realize that our environment can also be defined as other minds.[2] We spend our lives breathing in oxygen, whether we are aware of this or not. But—no less important—we also spend our lives interpreting and imagining minds, whether we are aware of this or not.

When people first hear about theory of mind, they often come away with two misconceptions. One results from imperfect terminology.[3] The word *theory* in "theory of mind" and the word *reading* in "mind reading" are potentially misleading. They seem to imply that we attribute states of mind intentionally and consciously: that is, *when we read minds, we know that we are reading them.*

Think again of my library example. It may have appeared from my description that I sat there droning silently to myself, "Hmm, I wonder why that woman in front of me is turning around and looking at those guys. Perhaps she appreciates this momentary diversion and wants to see its source. Or perhaps she wonders if she knows any of them." But of course it didn't happen that way. I wrote this event out for you as a sequence of fully articulated propositions because this is how we write and talk, but I certainly didn't experience it in such a neat, ordered, verbal fashion. I somehow "felt" all these possibilities almost at the same time, without intending to do so and without paying much attention to myself doing so.

It's difficult for us to appreciate just how much mind reading takes place on a level inaccessible to our consciousness. While our perceptual systems eagerly register information about people's bodies and their facial expressions, these systems do not necessarily make all of that information available to us for our conscious interpretation. Think of the functioning of "mirror neurons." Studies of imitation in monkeys and humans, made possible by advances in functional magnetic resonance imaging (fMRI) technology, have discovered a "neural mirror system that demonstrates an internal correlation between the representations of perceptual and motor functionalities."[4] What this means is that "an action is understood when its observation causes the motor system of the observer to 'resonate.'" So, for example, when you observe someone reaching for a cup, the "same population of neurons that control the execution of grasping movements becomes active in [your own] motor areas."[5] At least on some level your brain does not seem to distinguish between your doing something and another person's (whom you observe) doing it.[6] So you *understand* an action of another person—that is, you attribute a certain mental state to her—"She wants to grab that cup!"—because your mirror neurons are activated, but you have no control over or conscious awareness of their activation.

In fact, you don't even have to observe the action: the sound of an action (e.g., pressing a piano key) activates mirror neurons, too. Studies involving congenitally blind participants show that the "putative mirror neuron system can develop independently of vision." In this case, the system "projects the perceiver's own motor programs onto the sensory evidence of other people's actions rather than objectively mirroring the details of how the other has performed the action."[7]

Because the jury is still out on the role of mirror neurons and many aspects of that research remain controversial, my argument in this book does not depend on this research.[8] Still, with or without mirror neurons, we must have neural circuitry that is powerfully attuned to the presence, behavior, and emotional display of other members of our species. This attunement begins early (some form of it is already present in newborns), and it takes numerous nuanced forms as we grow into our environment. We are intensely aware of the body language and facial expressions of other people, even if the full extent and significance of such awareness escape our conscious notice.[9]

So when I was looking at that woman in the library turning around to look at the noisemakers, some of my mirror neurons must have been busy "being her," that is, perceiving the noise to her/my left and treating it as a disturbance, a welcome diversion, or a social opportunity. But then it also means that some of my mirror neurons must have been busy "being" those noisy people to the left. Otherwise I wouldn't have been able to infer that, by turning toward them, the woman was counting on their noticing her body language and interpreting it as meaning something about her attitude toward their actions. In other words, to the degree to which we—myself, that woman, and the people in that group—were aware of each other and were making sense of each other's behavior—our mirror neurons must have been involved in a three-way mutual modeling of our possible mental states.

I am having a surprisingly difficult time writing these things out. It makes me think that we don't have the vocabulary to explain how the functioning of mirror neuron systems underlies everyday mind attribution. It is so much easier to describe the workings of theory of mind the way I did in the beginning of this chapter. There I didn't try to model this mutually reflecting three-way process. Instead, I used the neat division between "me," "her," and "them" and simply attributed separate mental states to myself, to the woman, and to the group on her left.

For this *is* how we talk and write—and how I will have to talk throughout this book. To make the discussion of mental states manageable, we make it sound neatly isolated, evenly paced, intentional, self-conscious, and fully verbalized, as in, "I suspect that she is thinking that they don't realize that she is having a difficult time concentrating when they are whispering and laughing." Still, even if we have no choice but to talk about it this way, we should remember that this is not how our theory of mind really works. It's fast, messy, intuitive, not particularly conscious, and mostly not verbalized.

When We Read Minds, Do We Read Them Correctly? (The Second Misconception)

The second misconception about theory of mind is that reading minds means reading them *correctly*—a gussied-up version of plain old telepathy. In fact, nothing could be further from the truth (if anything, theory of mind is much more interesting than telepathy). Our mind-reading adaptations focus our interpretation of people's behavior on their mental states, but the interpretations themselves range from being completely wrong to only approximately accurate.

Here is one way to illustrate the difference between attributing mental states *constantly* and attributing them correctly. Foreign visitors or recent immigrants are bound to misinterpret certain gestures used by locals and hence misunderstand their intentions on such occasions. We treat these cases of miscommunication as striking and significant, but what's really striking and significant about them is the shared assumption, taken completely for granted by both newcomers and locals, that body language should be read in terms of underlying mental states. This assumption remains firmly in place no matter how many times communication fails as a result of misinterpreted gestures.

Consider the story of one such failure told by the literary critic Klarina Priborkin, born in Russia and now living in Israel, who remembers her family's first interaction with the Israeli immigration authorities upon entering the country:

> Losing patience after standing in line for several hours, the air-conditioning not working, some of the people who came with us on the plane decided to approach the authorities to ask how long this was going to take. When they returned, all they could report was that the clerk showed them a very strange gesture that somewhat resembled an offensive Russian gesture of "figa" [hand clenched in fist; thumb protruding between index finger and middle finger]. At least now we had something to pass the time, arguing about the meaning of the mysterious gesture. Later we learned that it is the Israeli gesture for "have patience!"[10]

Priborkin's story presents a strong argument against a universal language of gestures: a gesture deemed offensive by one culture may be considered conciliatory by another. Note, however, how the same story provides strong evidence for the universal adaptation for mind reading. Try answering the following questions without postulating some cognitive system (whatever you may want to call it) that irrevocably binds observable behavior to unobservable mental states:

First, what made the clerk assume that the newcomers would pay attention to her body language and thus notice the gesture she was making with her hand? After all, once they asked her the question, they could be looking at the nearby window as they waited for her reply. Second, what made the clerk assume that the newcomers would interpret her gesture as having a particular meaning? And why did the newcomers assume that this particular bodily movement of the clerk should have meaning, that is, be interpreted as indicative of a certain mental state? After all, they could have thought that her index finger was itching and she was unselfconsciously scratching it with her thumb.

For the sake of argument try answering these questions without bringing in theory of mind. You would have to propose that from our earliest childhood we are told by adults who surround us: "Pay attention to the bodies, my child. Note particularly the eyes, but do not neglect the mouth, either. A brow can tell you a lot about what the person is thinking. A wrinkled nose conveys much meaning. Hands are very important, but so can feet be, if properly attended to, in certain circumstances."

We don't often talk to children like this. True, when I read books with my toddler, I occasionally say to him things along the lines of "Mimi is unhappy because she lost her pet; she is crying; see that tear on her cheek?" or "See, Poombah is smiling because he likes what he did to the weaver." But comments such as these hardly add up to the incredibly powerful system of education that would have to be in place in every culture on Earth, from Abkhazia to Zuni, to bring about a universal tradition of interpreting body language as indicative of mental states.

And this fantastic system of education is what we would have to imagine—along with explaining just how it happens to come into existence in every single human society—if, for some reason, we don't want to postulate a cognitive adaptation for mind reading. In contrast, if we postulate such an adaptation, we say that our comments about Mimi's tears and Poombah's smiles merely *reinforce* rapidly maturing mind-reading predispositions of our preverbal audience rather than miraculously creating such predispositions from scratch. Indeed, developmental psychologists now study mindreading in seven-month-old infants,[11] and their research "has pointed to gradual, continuous, and universal stages in [theory of mind] development, that emerge in infancy and continue to progress during childhood and into early adolescence."[12]

To continue on a personal note: having emigrated from Russia to the United States in my early twenties and thus having had to consciously learn that the same gesture may mean very different things in the two countries, I always feel funny when I hear people emphasize such differences. Their arguments remind me that we tend to focus on exceptions and thus do not see the forest for the trees. Of course, I am quite aware of certain disparities between Russian and American body language; I learned about some of them through embarrassing personal experience. At the same time, however, I know that such disparities are completely dwarfed by what the two cultures have in common, that is, by the functioning of our theory of mind.

The very ability to notice cultural differences is evidence of theory of mind at work. For instance, cultures have different rules for emotional display. A "major task faced by the child in middle childhood is to learn the culture's display rules governing the conditions that are appropriate for the display of specific emotions, that is, situations in which the automatic urge to communicate the emotion currently experienced must be inhibited and either an alternative expression displayed or nothing revealed." But to observe and learn display rules one has to negotiate complex social stimuli—that is, read, misread, and reread minds constantly.[13]

In other words, we would have an extremely difficult time adjusting to new cultures (by figuring out, for example, local rules regarding display of emotions) were not all of our social interactions underwritten by the same evolved cognitive tendency to view observable behavior as caused by unobservable thoughts and feelings. The reason we can learn that in Russia "figa" means "when hell freezes over" while in Israel a somewhat similar gesture means "have patience" is that we have a strong cognitive predisposition to read gestures in terms of underlying mental states. Hence, to read minds constantly and unselfconsciously does not mean to read them *correctly* in any absolute sense. The most striking misreading of another's intentions is *still* mind reading—a fully realized exercise of our theory-of-mind adaptations.

So to come back to the cup-grasping example: your neural circuitry (whether represented by mirror neurons or some other dedicated systems) must underlie your understanding of my intention to grasp the cup, but you may never know, for example, if I reached for that cup because I was thirsty or because, for whatever reasons, I wanted you to think that I was thirsty. Thus any act of mind reading is fraught with possibilities for miscommunication and misinterpretation.

Greedy Mind Readers

In the rest of this chapter I argue that theory of mind is what makes our culture, as we know it, possible. It's a big claim, and it rests on two assumptions.

The first assumption is that our cognitive adaptations for mind reading are promiscuous, voracious, and proactive. They're always at work, stimulated either by actual or by imaginary interactions with other people. Encountering a body constitutes a powerful prompt for starting to attribute mental states. The body does not have to be real. Think of our reaction to people that we "meet" on canvases, on movie screens, or on the pages of a book. Although on some level we know that they are mere phantoms, our cognitive adaptations for mind reading still get in gear and start churning out interpretations of their thoughts and feelings.

We take all of this completely for granted, but pause for a moment and consider how strange this really is. "Caracalla's brow is knotted, and he abruptly turns his head over his left shoulder, as if he suspects danger from behind."[14] This sentence comes from a widely used art history textbook, *Gardner's Art through the Ages*. It describes a marble bust of the emperor Caracalla in the collection of the Metropolitan Museum of Art (fig. 1).

But Caracalla had been dead for eighteen hundred years! What is displayed at the Met is a carved chunk of marble! Still, what does our theory of mind care about such details? When confronted with that chunk of marble, we immediately interpret its bulges and concavities as indicative of a mental state, such as distrust. Alive or dead, marble or enamel, a human figure can't fail to provide grist for the mill of our insatiable mind-reading adaptations.

When I say "human figure," I mean both the full body and just the face. Faces, of course, are objects of our theory of mind's particular attention (though it seems that different cultures foster different strategies for "scanning" faces).[15] We are "addicted" to them from infancy.[16] As we grow older, we start seeing faces at the slightest suggestion: in clouds, in the random arrangement of dots, in chunks of marble.[17] We see them there not because of some general preference for facelike shapes but because we are foremost social beings, and facial expressions promise us access to the information most important for our well-being: other people's minds.[18] Whether or not they deliver on this promise is another question; I will return to it shortly.

To begin to appreciate the insatiable greediness of our mind-reading adaptations, it is useful to compare them to our adaptations for seeing. Because our species evolved to take in so much information about our environment visually, we simply cannot help seeing once we open our eyes in the morning (unless, of course, our visual system is severely damaged). The "predominance of sight" has had a profound influence on human culture: just think what a staggering range of daily practices directly depends on our ability to see.[19]

It's the same with mind reading, perhaps even more so: after all,

FIGURE 1. Portrait head of Caracalla (Emperor Marcus
Aurelius Antonius), ca. AD 217–30. Late Severan. Marble,
height 14¼ in.

blind people can't see, but they can attribute mental states.[20] As evolutionary psychologist Jesse M. Bering puts it, after a certain age people
"cannot turn off their mind-reading skills even if they want to. All human
actions are forevermore perceived to be the products of unobservable
mental states, and every behavior, therefore, is subject to intense sociocognitive scrutiny."[21] Hence, although we are far from grasping the full
extent to which our lives are structured by adaptations for mind reading,
we should expect cultural effects of those adaptations to prove just as
profound and far-ranging as the effects of the ability to see.

To get some idea of the scope of these effects, let's begin on a personal level. Talking to my friend and following her train of thoughts offers the most immediate input for my theory of mind. So, too, when she is away, does imagining what she might be thinking at this moment. So, too, if she dies, does imagining what she would have thought on such and such occasion.[22]

I want more, however. I want to hear stories about what other people did and what they looked like when they did it so that I can imagine what they thought and felt at those times. Those people can be members of my family or complete strangers or people that never existed. They don't even have to be human: androids, talking animals, dancing candelabras, and twinkling stars will do. I can listen to such stories; I can read them; I can hear them sung; I can watch them danced or mimed or projected on a flat surface; I can look at them carved into stone, painted on walls, or reproduced in art books. Because I want to see bodies in action so that I can think about their intentions, sometimes I make up those stories myself in whatever way I can: whether painting, dancing, singing, or writing. In my particular case this may involve writing about what fictional characters and their creators might have meant when they did this or said that, as well as about what other scholars, dead or alive, might have said or did say about this or that. Literary critics make a living by reading and misreading minds.

I am talking about myself here. Now think: if other people have the same need to process mental states, what kind of culture must emerge in response to this need? This culture has to continuously feed this need, yet it will never be able to fully satisfy it since new mind-reading cravings arise all the time. It is a *culture of greedy mind readers* (bound to become even greedier with the advent of a media-saturated society, as new modes of storytelling seem to appear constantly). A case in point: five years ago I could not foresee that today I would need to read a particular blog regularly. Back then I didn't even know what blogging was. And now I am addicted to this blogger's way of thinking: I crave my daily fix of her mental states.

Here are some phenomena that one might encounter in a culture of greedy mind readers: stories that depict people's response to their perception of other minds (such as novels); arrangements that let us read mental states into sequences of movements set to music (such as ballet); specially designated social spaces in which we can appreciate the gap between what people feel and what they would feel had they known as much, or as little, about their situation as we do (such as theater); events during which numerous physical bodies form complex patterns guided by the shared understanding of intentions (such as team sports); and artifacts that coordinate text and images so that the information about people's feelings that we get from looking at their body language elaborates, contradicts, or otherwise complicates the verbal descriptions of their feelings (such as graphic narratives).

I am obviously talking about cultures that I am most familiar with. Had I been born in Bali and moved first to Java and then to South Sulawesi instead of being born in Russia and moving first to Latvia and then to the United States, my examples might have featured more prominently theater-performed puppet shows, forms of dancing that tell stories about ancient Buddhist kingdoms, wood carvings, or funereal rites.[23] There is no predicting what forms cultural phenomena that feed our theory of mind will take in a concrete historical moment in a particular society. We can predict, however, that no cultural form will endure unless it lets us attribute mental states to somebody or something.

Imagine the impossible: our theory of mind is switched off. How many cultural institutions that let us read minds into behavior would survive? Who would attend bullfighting, pantomime, basketball games, opera, finger-shadows theater, or tightrope walking? If you doubt that tightrope walking engages our theory of mind, consider this: we *know* that the performer *does not want* to die and that she *knows* that what she is doing is dangerous; moreover, she *knows* that we *know* that she *knows* that what she is doing s dangerous. That's why a performer sometimes pretends to slip and nearly fall down, eliciting a collective gasp from her audience below. She is playing with our minds, making us imagine what she must feel as she narrowly escapes death. Take this unconscious attribution of mental states out of the act of tightrope walking and see how interesting it remains. In fact, drained of all mind-reading, tightrope walking is exactly as interesting as a Whee-lo toy rolling back and forth on its magnetic axle.

Just so, watching a basketball game without attributing intentions to players is as enticing as watching falling snowflakes—both are random movements, fascinating for about two minutes, and then your mind wanders off. Opera is a pain: bodies moving haphazardly across a stage, bursting into song at random intervals. Finger shadows: why is that woman moving her hands this way? With our theory of mind intact, we say it's because she *wants* to imitate the movement of a dog's tail—she *wants* to *amuse* us. But without theory of mind her random twitching and twisting of hands seems incomprehensible, unsettling, perhaps threatening.

Now think about the fate of social, political, and economic networks built around a variety of orally transmitted narratives, public rituals, novels, movies, plays, cartoons, news reports, sporting events, online discussions, and, more fundamentally, our everyday conversations about people's plans, thoughts, and feelings. These networks would crumble because they are only sustained by our ability and need to read mental states into behavior. And once the networks of the culture of greedy mind readers are gone, what's left?

The Best and the Worst

Here is the second of the two assumptions behind my claim that theory of mind makes human culture possible: bodies are simultaneously the best and worst source of information about people's thoughts and feelings.

That is, we *perceive* bodies as both the best and the worst. On the one hand, we put tremendous value on the information about people's mental states that we glean from their body language. On the other hand, we are always ready to turn about and treat this information as particularly unreliable. This paradoxical double perspective is fundamental and inescapable; it informs all of our social life and cultural representations.

To appreciate the power of this double perspective, imagine that right now you and I are talking face to face. Let's say you are trying to convince me of something. As we go on, you know that I am not merely listening to your words but also paying attention to your face, movements, and appearance. That is, you generally can't know what particular grin or shrug or shift in affect I notice and consider significant at a given moment; indeed, I don't know either. Still, our long evolutionary history as a social species—expressed in our cognitive adaptations for mind reading—ensures that you intuitively expect me to read your body as indicative of your thoughts, desires, and intentions and that my reading of your body will be crucial for the outcome of our communication.[24] Moreover, the same long evolutionary history ensures that I intuitively know that you expect me to read your body in this fashion. That is, I know that you will *perform* your body language, though not necessarily consciously or intentionally, to influence my perception of your mental states.

This means that I have to constantly negotiate between trusting this or that aspect of your observable behavior more than another. Were I to put this negotiation into words—which will sound funny because we do not consciously articulate it to ourselves this way—it might go as follows: "Did she smile just now because she liked what I said or because she wanted me to think that she liked what I said, or because she was thinking of how well she handled an argument yesterday, or was she thinking of something altogether unrelated?"

Thus, we treat with caution the information about the person's state of mind inferred from her observable behavior precisely because we can't help treating observable behavior as a highly valuable source of information about her mind—*and we both know it.* Because we read intentions into bodies throughout our evolution as a social species, we are now stuck, for better or for worse, with cognitive adaptations that forcefully focus our attention on the body.

Nor would we want to completely distrust the body—our far-from-perfect readings of each other get us through the day. Still, as we automatically interpret each other's observable behavior in terms of underlying mental states, on some level we keep active the hypothesis that the observable behavior is misleading. (Note, too, that it does not have to be intentionally misleading. If I meet a person whose natural expression is a frown, I may incorrectly assume that he does not like me. The body may misrepresent the mind.)

From Private Mind Reading to Cultural Arms Race

So we are in a bind. We have the hungry theory of mind that needs constant input in the form of observable behavior indicative of unobservable mental states. And we have the body on which our theory of mind evolved to focus so that it can get such input. And that body, by virtue of being the object of our theory of mind's obsessive attention, is a *tremendously valuable and, as such, potentially misleading* source of information about the person's mental state.

I make this argument based on research in different branches of psychology, specifically, evolutionary psychology, developmental psychology, and cognitive neuroscience. When I turn to other academic disciplines, however, such as sociology and literary criticism, I see some exciting overlap of ideas. For instance, in 1969 the sociologist Erving Goffman observed that in human communication, increased reliability leads to increased unreliability:

> The more the observer relies on seeking out foolproof cues, the more vulnerable he should appreciate he has become to the exploitation of his efforts. For, after all, the most reliance-inspiring conduct on the subject's part is exactly the conduct that it would be most advantageous for him to fake if he wanted to hoodwink the observer. The very fact that the observer finds himself looking to a particular bit of evidence as an incorruptible check on what is or might be corrupted is the very reason why he should be suspicious of this evidence; *for the best evidence for him is also the best evidence for the subject to tamper with.*[25]

Goffman's larger argument about the vulnerability of incorruptible evidence works perfectly with my more specific argument about body language and mental states. To the extent to which our mind-reading adaptations make us see bodies as providing "foolproof" cues to thoughts and feelings, we remain vulnerable to convincingly faked body language.

Similarly, scholars in literary and cultural studies have commented extensively on the protean nature of the body as they have sought to expand the concept of performance beyond the theatrical stage to a broad range of everyday practices.[26] Research on theory of mind lends strong support to their insights. Because we are drawn to each other's bodies in our quest to figure out each other's thoughts and intentions, we end up *performing our bodies* (to adapt a term from cultural studies) to shape other people's perceptions of our mental states.[27]

Again, this may seem like a mere description of private interpersonal dynamics, but let's expand it to our culture as a whole. It turns out that a broad variety of daily practices reflect the dual position of the body as a valuable yet unreliable source of information about the mind. For example, our social infrastructure seems to be chock-full of devices designed to bypass the body in reading a person's intentions. We use blood and hair samples, credit and medical histories, and fingerprinting to avoid the situation in which we have to make an important decision based on information provided solely by the person's observable behavior.[28]

Some of these devices work better than others, but none are perfect. We may not yet be living in the future depicted in the movie *Gattaca* (1997), whose protagonist, Vincent, fakes his blood and hair samples to deceive others about his intentions. (To be precise, he deceives others about his genetic identity, but, in Vincent's world, genetic identity is synonymous with intentions: it is supposed to determine what a person should dare to aspire to and what he should think of his place in society and relations with others.) Still, that sci-fi moment does capture an important sociocognitive feature of our world. There is a constant arms race going on between cultural institutions trying to claim some aspects of the body as essential, unfakeable, and intentionality-free and individuals finding ways to *perform* even those seemingly unperformable aspects of the body.[29]

A memory from a real-life dystopia: in the 1980s I learned that I would be required to take an eye exam in order to go to college. I don't remember the details, but being very nearsighted in Soviet Russia barred one from a number of activities, including getting a driver's license, participating in certain sports, giving birth without surgical intervention (it was thought that one's retina could detach during labor), and, apparently, studying at Moscow State University (MGU). At least that's what they said in my home town in the Ural Mountains when I started putting together my college application. And, no, I didn't plan to be a fighter pilot; I wanted to major in journalism.

I had to find a way to fake an acceptable level of nearsightedness during the eye exam. I procured—I don't remember how—a copy of the eye chart and learned it by heart using mnemonics. That is, I composed a short verse, with two words in the first line, three words in the second, five in the third, five in the fourth, etc.—all beginning with the letters on the chart. I still know that doggerel by heart, so if there is a chance that Russian eye charts haven't kept up with other changes in that country, such as the prices of oil and law enforcement officers, I could still, perhaps, pass there as someone worthy of driving a car, playing tennis, and going to college.

I did get into the formidable MGU and ended up receiving an education that was mostly a joke. So my plan must have worked, to some extent, even though I don't remember any details from my visit to the eye doctor. What I do remember, however—and that's why I am telling this story—is how determined I was to fake that seemingly unfakeable, seemingly essential physical attribute: good vision. If they wanted to decide my future based on their reading of my body, I would make them see that body in a way that suited my intentions and not their arbitrary assumptions about what a nearsighted person could or couldn't do.

So for every effort to read a body for incontrovertible evidence of something essential about a person, there will be a countereffort directed at manipulating the mind that is supposed to be doing that reading. The countereffort may fail but not for want of trying. The arms race between those who want to fix the meaning of the body and those who want to influence the mind that fixes the meaning may take different forms in different historical contexts, but it seems to be an unavoidable feature of a culture of greedy mind readers.[30]

Conclusion: To Know and Know Not

We read minds all the time yet remain open to the possibility that our readings are wrong. With such a peculiar setup in place, what should we expect from our cultural representations? Of course, this big question cannot be answered in one book. But as a starting point, let us consider step-by-step what it means to live in a world in which we know, and at the same time don't know, what other people are thinking.[31]

First, we *assume* that there must be a mental state behind an observable behavior. Say you see somebody jumping up in the middle of a meeting. Try making sense of his action without talking about his presumed mental state; for example, he had an idea; he remembered something suddenly; he wanted to see how high he could jump; he felt something sharp on the seat beneath him; he saw a snake and was terrified; he wanted to determine whether everybody was awake.

Our belief that there must be a mental state behind a behavior is itself a cognitive artifact that reflects the way we perceive people. The question of whether my colleague over there *truly and really* had some thought, feeling, or emotion that prompted him to jump is relatively irrelevant.[32] What is relevant is that for us, that jump signals an underlying mental state.[33]

Second, even though we know that there must be a mental state behind a behavior, we don't really know what that state is. There is always a possibility that something else is going on behind even the most seemingly transparent behavior. We can remember situations when our thoughts did not fit the circumstances, and no observable behavior could reveal them to people around us—or so we hope. On these occasions we say to ourselves, "Thank God, we can't read each other's minds, so that they have no way of knowing what is going through my head."

Third, even though we can't really know what other people are thinking, we conduct our daily lives on the assumption that we do, more or less. To borrow from a related discussion by the cognitive literary critic Ellen Spolsky, our everyday mind-attributions are "good enough."[34] Obviously, I can't be entirely sure what that woman is really thinking as she strides determinedly toward that particular weightlifting machine, but it has served me well in the past and is likely to serve me well in the future to assume that she wants to use it right away, which means that for the next five minutes I'd better turn to a different machine. Such rough-and-ready interpretations get us through the day. To quote cognitive evolutionary anthropologist Dan Sperber, in "our everyday striving to understand others, we make do with partial and speculative interpretations (the more different from us the others, the more speculative the interpretation). For all their incompleteness and uncertainty, these interpretations help us— us individuals, us peoples—to live with one another."[35]

Were we to stop and try to figure out what the people around us are really thinking, we would become socially incapacitated, overwhelmed with possible interpretations, and unable to commit to any course of action. Perhaps the reason that we even notice our moments of "Thank God, we can't read each other's minds!" is that they stand out amid our daily unreflective mind attribution. They interrupt its course. They force us to juxtapose a *good enough* mind attribution—that is, what people are likely to be thinking in such a situation—with an *exact* and unexpected mind attribution: what I really thought in that situation.

Fourth, because we go around knowing that there must be a mental state behind a behavior, and because we don't really know what that state is, even as we act as if we know, cultural representations exploit this precarious state of knowing and not knowing. A writer makes us think that a protagonist goes into a catatonic stupor because she is distraught by the news that her husband was killed in an accident only to reveal later that she could not move because she was overcome with happiness over her sudden freedom. An artist paints a smiling woman but gives us no context in which to interpret that smile and thus leaves us forever intrigued about her thoughts. A stand-up comedian exploits various contexts in which people in his culture are strongly expected to be sincere—using the body language appropriate to each of these heartfelt occasions—only to drive his audience to such a state of bewilderment and skepticism that when he actually dies from kidney failure caused by cancer, they don't believe it.[36] The more we look for the "true" mind in the body, the less we can hope to find, yet every screen, every stage, every page offers us new ways of looking.

TWO

In which a new concept is introduced / a phobia is revealed / a four-letter word makes a bold-faced appearance (but the French take the blame) / Frederick Wentworth betrays himself / Elizabeth Bennet rejects Mr. Darcy / Bridget Jones triumphs over a rival / Tom Jones can't see what's in front of his eyes / and the author admits that she has no clue what her nearest and dearest are thinking.

I Know What You're Thinking, Mr. Darcy!

Embodied Transparency

Jane Austen's novel *Persuasion* (1816) tells the story of a woman who, unmarried and unhappy at twenty-seven, suddenly finds herself thrust into the company of a man whom she has loved but was persuaded to give up eight years ago. The objections that her friends had to him (poor, lacking social connections) are moot today, for he has made a brilliant and lucrative career in the Royal Navy. But it's too late. He is not interested in her anymore, looking instead for someone with courage and conviction, someone whose opinion won't be swayed by shortsighted well-wishers. Perhaps someone younger, too.

Or so Anne Elliot thinks. Dispirited as she is, she is only too ready to read Captain Wentworth's behavior toward her as mere polite indifference. Whenever they see each other, which usually happens at the house of Anne's relatives, the Musgroves, he comes across as happy, satisfied with himself, and as civil toward her as he would be toward any other old acquaintance:

It was a merry, joyous party, and no one seemed in higher spirits than Captain Wentworth. She felt that he had every thing to elevate him, which general attention and deference, and especially the attention of all the young women could do. . . . If he were a little spoilt by such universal, such eager admiration, who could wonder?

These were some of the thoughts which occupied Anne, while her fingers were mechanically at work, proceeding for half an hour together, equally without error, and without consciousness. Once she felt that he was looking at herself, observing her altered features, perhaps, trying to trace in them the ruins of the face which had once charmed him; and once she knew that he must have spoken of her: she was hardly aware of it till she heard the answer; but then she was sure of his having asked his partner whether Miss Elliot never danced? The answer was, "Oh! no, never; she has quite given up dancing. She had rather play. She is never tired of playing." Once, too, he spoke to her. She had left the instrument on the dancing being over, and he had sat down to try to make out an air which he wished to give the Miss Musgroves an idea of. Unintentionally she returned to that part of the room; he saw her, and instantly rising, said, with studied politeness—

"I beg your pardon, madam, this is your seat"; and though she immediately drew back with a decided negative, he was not to be induced to sit down again.

Anne did not wish for more of such looks and speeches. His cold politeness, his ceremonious grace, were worse than any thing.[1]

There is a moment during the party, however, that contrasts sharply with Frederick Wentworth's general affect of vague politeness and complacency, a moment in which Anne feels that she knows exactly what he is thinking. At one point, Mrs. Musgrove speaks to him about her late son, Richard, who used to serve under his command and, though an unpromising and careless young man, apparently behaved somewhat more conscientiously when supervised by Captain Wentworth:

"Poor dear fellow!" continued Mrs. Musgrove; "he was grown so steady, and such an excellent correspondent, while he was under your care! Ah! it would have been a happy thing, if he had never left you. I assure you, Captain Wentworth, we are very sorry he ever left you."

There was a *momentary* expression in Captain Wentworth's face at this speech, a certain glance of his bright eye, and curl of his handsome mouth, which convinced Anne, that instead of sharing in Mrs. Musgrove's kind wishes, as to her son, he had probably been at some pains to get rid of him; but it was too *transient* an indulgence of self-amusement to be detected by any who understood him less than herself; *in another moment* he was perfectly collected and serious, and, *almost instantly afterwards* coming up to the sofa, on which she and Mrs. Musgrove were sitting, took a place by the latter, and entered into conversation with her, in a low voice, about her son, doing it with so much sympathy and natural grace, as shewed the kindest consideration for all that was real and unabsurd in the parent's feelings. (63–64, emphasis added)

There is a striking contrast between the Frederick of the rest of the chapter, whose thoughts and feelings can only be guessed at, and the Frederick of this passage, whose involuntary look and smile render him transparent to Anne. She gathers that Frederick didn't think highly of Dick Musgrove and that he wants to conceal that from the young man's mother. This moment of perfect access is over almost immediately—in the long quote above I italicized the phrases that emphasize just how transient it is—yet it's there for Anne to notice with a pang of old intimacy.

I came up with a special term to describe the moments in fictional narratives when characters' body language involuntarily betrays their feelings, particularly if they want to conceal them from others, as Frederick Wentworth does. I call it *embodied transparency* and believe that the pleasure that we as readers derive from such moments is best explained by thinking about what they do to our theory of mind. (What they do to the characters inside the story is a different matter; many characters don't even notice them.) Instances of embodied transparency offer us something that we hold at a premium in our everyday life and never get much of: the experience of perfect access to other people's minds in complex social situations. As such, they must be immensely flattering to our theory-of-mind adaptations, which evolved to read minds through bodies but have to constantly contend with the possibility of misreading and resulting social failure.[2]

Embodied transparency is but one of many ways in which fiction engages our mind-reading adaptations. As I argued in *Why We Read Fiction,* theory of mind makes fiction, as we know it, possible.[3] To read a work of fiction is to attribute mental states to fictional characters, to the writer, and to oneself—the exact balance and configuration of these three types of attribution depending on the genre and style of the specific piece.[4]

What my earlier broader argument about fiction and theory of mind has in common with my present argument about embodied transparency is an emphasis on the social. Theory of mind evolved to track mental states involved in real-life social interactions, but on some level our mind-reading adaptations do not distinguish between the mental states of real people and those of fictional characters.[5] Fictional narratives feed our hungry theory of mind, giving us carefully crafted, emotionally and aesthetically compelling social contexts shot through with mind-reading opportunities. The pleasure afforded by following minds on the page is thus to a significant degree a social pleasure—an illusory but satisfying confirmation that we remain competent players in the social game that is our life.

But this is how what I do here is different from what I did in *Why We Read Fiction.* In that book I showed that modernist fiction, novels featuring unreliable narrators, and detective stories play with our mind-reading adaptations by keeping them off-balance. That is, they make us weigh and reweigh the truth-value of information that we glean by following mental states that are strategically embedded within other mental states. For instance, we gather that character A wants character B to think that character C has betrayed B, yet we are not sure about character A's motivation and thus don't know if she is honest with B. Thus *Why We Read Fiction* focuses on the mind-reading *uncertainty* that the manipulation of mental states induces in us: the characters manipulate mental states of each other, the narrator manipulates mental states of the reader, and so forth.

In contrast, here I look at mind-reading *certainty* and body language as the path to that certainty. This means that while in *Why We Read Fiction* body language is dealt with only incidentally, here it assumes center stage. This focus on the body limits my argument in one way and opens it up in another. It limits what I can say about prose fiction because fiction writers use embodied transparency sparingly. Austen, for instance, would rather rely on the prerogative of the omniscient narrator to tell the reader what her character is thinking (as in the case of Anne, above), or else would make us grope in the dark along with one character, trying to guess what's on the other character's mind (as we are groping along with Anne, trying to guess if Captain Wentworth still cares about her).

But if embodied transparency is relatively rare in prose fiction, it's abundantly present in visual media: movies, musicals, paintings, and reality shows. So focusing on the body as the direct pathway to mind opens up new ways of looking at a variety of cultural phenomena. What's interesting about this wider view is how many "rules" for depicting embodied transparency turn out to translate across genres. In the rest of this chapter I will talk about these rules in fiction, setting the stage for my discussion of visual representations in chapters 5 through 10.

How Is Embodied Transparency in Fiction Different from Embodied Transparency in Real Life?

When I claim that fictional moments of embodied transparency treat us with the direct access to other people's minds that we never get enough of in real life, I don't mean that in real life we never get to intuit what other people think or feel based on their observable body language. Of course we do! (Or at least we think we do, which is the same.) What I mean, rather, is that in real life we almost never encounter the combination of direct access and social complexity that fiction offers us on a regular basis.

Think about it this way. In real life the correlation between social complexity and transparency is negative. The more socially complex the situation is—that is, the more mental states we need to follow in order to grasp it—the more possibilities there are for misinterpreting what seems to be transparent body language. In fiction the correlation is positive. Writers build extremely involved social situations to bring characters to a point at which their bodies fully reveal their minds.

Hence in a novel it is because character A thinks that she knows what character W may think, both about character R and about an appropriate emotional response to character M's feelings about character R, that readers believe that they know what character W is thinking when he glances up and curls "his handsome mouth" in a certain way. In contrast, in real life we may observe person W glancing up and half-smiling, and we may even know how he feels about M's attitude toward R, but there are still so many opportunities for misreading W's body language on this particular occasion that we would be either naive or delusional to think that we know exactly what is going through his mind. We may think that he is amused by what M says about R, while in reality he may be thinking of how much his dog loves its new toy.

So when we look for cases of embodied transparency in our everyday life, more often than not we come up with socially impoverished situations, that is, situations that don't require attribution of mental states embedded within other mental states embedded within other mental states (as in "A wants B to think that C didn't want B to know about X"). I briskly pass the service desk while exiting the gym but then realize that I forgot my bicycle helmet back in the locker. I turn around and approach the desk with an ingratiating smile, hoping that they'll just let me in and that I won't have to fish for the gym ID in my backpack. The girl at the service desk must have been watching me exit and then stop suddenly and turn around, because even before I begin to explain, she nods, says, "Forgot something?" and lets me in. For the last ten seconds my body language has been perfectly transparent to her, but this instance of embodied transparency involves only two mental states—she *knows* that I *want to retrieve* a forgotten object—so it is boring. Nothing to write novels about.

Not to forget about such obvious physiological examples of real-life embodied transparency as sneezing, having erections and orgasms,[6] burping, passing gas, jerking away one's hand when accidentally touching a hot stove, and so forth. Once again, none of these are interesting unless we start adding more mental states to them, but as we do, transparency evaporates.

Take orgasm. On the one hand erotic love seems to create compelling contexts for embodied transparency (indeed, the term *intimacy* itself, as it is currently used in our culture, appears to reflect this ideal of perfect access). On the other hand orgasm can be faked, and, if it isn't, it can still be "performed," that is, rendered more visually and aurally expressive. Look how it changes the mind-reading dynamics of the situation.

A person who consciously intensifies the performance of her orgasm is attempting to manipulate the mind of her partner—that is, she *wants* him or her to *believe* that she *feels* X. Conversely, the partner, who is aware of the possibility of such a performance, may wonder if the person who is having an orgasm really "means it" or if she wants him or her to believe that she feels X. Argh! The complexity of the situation increases at the expense of transparency—a zero-sum game that haunts real life but can be escaped in fiction.

As I keep searching for other instances of everyday embodied transparency—besides achieving wordless understanding at service desks, jerking a hand away from a hot stove, sneezing, and having honest-to-goodness orgasms—I turn to babies and pets. Babies certainly represent a fascinating example of the trade-off between social complexity and transparency. Although even a five-month-old can manipulate a parent by crying, not because he is hungry or colicky but because he wants to be picked up and cuddled (thus essentially "performing" pain and distress for his receptive audience), babies are still quite transparent much of the time. When a one-year-old reaches for a ball, that ball is what she wants. When she embarks toward it across the room, her single-mindedness of purpose is a sight to behold. When a fifteen-month-old insists on being read this but not that book, he does it not because he wants to impress his parents in a certain way—that is, not because he *wants* his parents to *think* that he *thinks* X—but because this book is what he absolutely wants to be read right now. When he smiles, he means it, and he certainly means it when he laughs.

Obviously, there are many reasons, ranging from the effects of oxytocin to cultural traditions, why people like babies, but I believe that this capacity for embodied transparency contributes to the delight we take in them. Our theory-of-mind adaptations seem to go into high gear when we sense a possibility of witnessing a real—that is, not put-on or performed—emotion; and babies tend to rivet our attention with the transparency of their emotions practically all the time.

Think, too, what underlies the compliment that we pay to a person when we say that he or she is "childlike." When we do so, we typically refer to the immediacy and freshness of that individual's emotional responses; we appreciate the candor, the lack of pretense. But what underlies this compliment (if we look at it squarely from a mind-reading perspective, which is not a perspective we'd be consciously aware of) is that we feel that we can read their emotional responses, and we like that.

Pets can offer us real-life embodied transparency, dogs perhaps more so than cats (though cat owners may disagree). Certain lizards are delightful pets because they change their skin color depending on their emotions.

As these last examples show, I have reached the bottom of a barrel that was not particularly deep to begin with. Real life is either stingy or not very socially exciting when it comes to embodied transparency. If it were different—that is, if we could regularly come upon socially complex embodied transparency in real life—fictional stories featuring transparency would have had no chance. We would have remained glued to watching other people (and ourselves) interacting with each other and betraying their (and our own) feelings in wonderfully informative ways. But as it is, watchful as we may be, embodied transparency that does come our way can't hold a candle to that available in books, movies, and plays.

But if real life is a tightfisted bore, what do fiction and, especially, movies—reservoirs of titillating, socially rich transparency—do to us? Does consuming embodied transparency on the page, onstage, and onscreen sharpen our appetite for it in our everyday life? Do we start perceiving people around us as more transparent than they are? Or do we get addicted to shows and stories that offer us a steady supply of readable bodies?

I think the latter is certainly the case, and the former might be the case; but I wouldn't know how to test it. An experiment demonstrating that after watching a lot of reality TV a person is more prone to believe that people around her often inadvertently reveal their feelings may not say much about the long-term social effects of watching such shows.

One reason that embodied transparency in a complex social situation exists only in fiction, movies, and television shows may have to do with the way our consciousness operates. Fictional representations can successfully create the illusion of transparency because fictional minds are, in principle, fully knowable while, in real life, the very notion of fully knowing one's mind or someone else's mind is problematic.

What does it mean to "really" know your mind? As evolutionary psychologist Robert Kurzban shows in his remarkable recent book, *Why*

Everyone (Else) Is a Hypocrite, it means to subscribe, without even being aware of it, to a version of the notorious Cartesian theater (a concept introduced by the philosopher Daniel Dennett): the homuncular "you" at the center of your consciousness observing what your "mind" is up to.[7] But there is no theater in the brain and no "you" in the best seat of the house, deciding which of the feelings currently on display are "real" and which are only approximately real or downright fake.

It may be even worse than that. It seems that parts of your brain don't communicate with each other—the "brain might be designed to keep certain parts of information away from other parts"—and they "can simultaneously hold different, mutually contradictory views."[8] This means that consciousness is gappy and discontinuous all the way through[9] and not amenable to any kind of full introspection, or "transparency," either from the inside or the outside. (Once more, we are talking about complex social situations; a person screaming desperately at the sight of the murderer with an ax is transparent both from the inside and the outside.)

But if one can hold several contradictory views simultaneously, what can body language presumably betraying "real" feelings in fact betray? If we assume that the inadvertently revealed feelings exist on a conscious level—that is, if the person is aware of these feelings—we are in trouble because what she is aware of is not more "real" than what she is not aware of. And if we assume that the inadvertently revealed feelings are those that the person is not in the least aware of, how can we say that these are her "real" feelings?

What possessed her to do that? What possessed me to do this? I don't know. At times we face the deeply uncomfortable realization that we may never fully comprehend the mental states behind this or that action of our own or of other people that seems so strange, so meaningless, so out-of-character. "If you are like me," Kurzban writes, "you have often—and quite honestly—answered the question 'Why did you do that?' with 'I have absolutely no idea.'"[10] Our theory of mind is not terribly well equipped to deal with a discontinuous consciousness (i.e., a consciousness defined by a lack of communication between different parts of the brain) and with the behavior that comes out of this discontinuity.

But, equipped or not, theory of mind never quits. So we keep coming up with explanations—"surely, I must have wanted X when I did that!" "Oh, he must have been thinking Z when he did this!"—even as the explanations ring hollow. We may turn to essentialism: "He acted this way because this is just the way he is." But essentialist thinking is not very satisfying either because it seeks to shut down our theory of mind, and *good luck with that!* In the midst of our best essentialist effort we still keep wondering if there could be some motivation, desire, or thought that we haven't considered and that may yet explain inexplicable behavior.[11]

Not so with fictional characters. Unpredictable as a character can be and discontinuous as her consciousness may seem, it's hard for us to let go of the intuition that someone, somewhere, under certain circumstances, has, or used to have, a privileged insight into that consciousness.[12] When a story is marked off as fictional (which is a complex cultural process that exploits our cognitive adaptations for source-monitoring),[13] it's perceived as a story with a mind behind it: the mind of its author, even if the author is anonymous or unknown.[14] This makes the minds of characters within the story knowable, at least in principle. So if the author is inclined to work in some moments of embodied transparency, this transparency can be believable in a way that real-life transparency can't be.

(The real-life analogy to this mind-knowing author is, of course, God. If you can't fathom why someone did something, you can still throw a bone to your hungry theory of mind by saying, "Only God knows what he was thinking when he did that!" But if your personal universe admits no God, your theory of mind is out of luck.)[15]

Three Rules for Embodied Transparency in Prose Fiction

There seem to be three "rules" for constructing moments of embodied transparency in prose fiction. The first rule is *contrasts:* an author has to build up a context in which the character's transparency stands out sharply against the relative lack of transparency of other characters or of the same character a moment ago or a moment after. The second rule is *transience:* to be believable, instances of transparency must be brief. The third rule is *restraint:* more often than not, characters struggle to conceal their feelings and by doing so become transparent.

These rules are not absolute. As I will shortly show, certain genres, such as fairy tales and stories featuring unreliable narrators, violate them routinely. Yet they are consistent enough for me to suggest that when we encounter what seems to be an instance of embodied transparency in fiction, these rules are worth checking for. Whether followed or violated, they come through in interestingly idiosyncratic ways.

What are the origins of these rules? The need for contrasts might be the fictional narrative's holdover from visual cognition. Any art that appeals to the eye must cultivate gradients. As art historian Ernst Gombrich has observed, "Even newly hatched chickens classify their impressions according to relationships."[16] So, perhaps, to the extent to which we visualize what we read, thinking of characters in relative terms (e.g., he looks happier than his friend; she is taller than her sisters; his face is more expressive than hers) enhances the image's sensory appeal.[17]

The emphasis on transience seems to be directly related to our daily practices of mind reading. Imagine a context in which a person seems momentarily transparent to the people around her, and ask yourself how long this transparency can last before it turns into a performance.

Say that you know me well and know that I am deathly afraid of mice. At some point in a public place, somebody points to the ground close to my feet and exclaims, "Mouse!" Most likely, I will let out a shriek, and, for that second, you can be quite sure that you know what I feel: fear, disgust, and the wish to get further away from the mouse, wherever it is. Just then I will be transparent—that is, as transparent as people can be in real life. But if I continue shrieking or looking terrified for more than a second, what started out as transparency is likely to have changed into performance. (Unless my shrieks are an indication of drunkenness or mental breakdown—in which case I am still transparent—but altered mental states form a category of their own.) That is, it's possible that by continuing to shriek I am trying to impress others in a certain way: perhaps I want to come across as weak and feminine—if my culture considers displays of weakness in women attractive—or simply want to hold everybody's attention. But whatever my motivation, you're not sure what it is, which means that I am not transparent anymore.

It is the same with fictional characters. The longer the character seems to be transparent, the more likely it is that, knowing that other characters are reading her body language as indicative of her mental states, she is trying to manipulate their thinking. There is a very small window of opportunity during which a sober, sane character, *not* in an acute life-threatening situation, can be seen as transparent without lapsing into affectation and thus becoming opaque again.

The other side of the need for transience is ethics. If a character is forced into being transparent while another character is watching, and all this goes on for too long, the observer will soon be perceived as sadistic. And if you are a writer and you want your hero or heroine to remain sympathetic, you don't put them in a situation in which they begin to enjoy the spectacle of someone else's transparency. (I will discuss shortly the first marriage proposal scene from *Pride and Prejudice,* in which Austen makes sure to tell us that Elizabeth feels "dreadful" as she watches Mr. Darcy's struggle to conceal his anger and shock, even though that moment of struggle, and hence transparency, must have been quite brief.)

This reflects in interesting ways what happens in real life when bodies betray people's feelings in full view of others—when, for instance, a person bursts into tears or blushes violently. Although people respond to such situations in a variety of ways, depending on the context, one recognizable reaction is to avert one's gaze or, at least, to make a show of averting one's gaze. What this reaction seems to imply is that people are put at a social disadvantage when their faces "leak" emotions against their wills, so it is wrong of others to use such moments of weakness to learn something about the "leaking" person's real feelings and exploit that knowledge in subsequent dealings with them.

The phenomenon of leakage has been much debated by psychologists. Some think that it doesn't really exist. That is, people certainly exhibit involuntary body language, including bursting into tears and blushing, but this language may not be as informative about "true" mental states as it is made out to be. (This view is compatible with but not identical to Kurzban's broader questioning of the concept of "true" feelings.) Thus psychologist Alan J. Fridlund argues that, from the evolutionary perspective, it makes no sense to expect that natural selection would favor a system of signaling that provides information "detrimental to the signaler" and that "displays generally should not necessarily occur when one is emotional at all, but when they will do the most good for the displayer."[18]

In other words social rules against staring at people when they involuntarily display their feelings are based on our biased view of the body as a uniquely valuable source of information about the mind. This view may reflect our evolutionary history as a mind-reading species (i.e., we read bodies for information about mental states for hundreds of thousands of years, and we are pretty much stuck with doing this) but may overestimate the actual ability of body language to provide direct access to mental states.

So when writers keep embodied transparency brief so as not to let other characters observe it for long, they follow our intuition that bodies that leak feelings are socially vulnerable. This intuition may be correct only to some limited extent—social vulnerability can certainly be displayed strategically so as to "do the most good for the displayer"—but fictional narratives tend to abide by this conventional view. As long as the body is perceived as the privileged pathway to a person's mind, authors will not allow a sympathetic character to dwell on someone else's transparency to her heart's content. And she would not entertain such a heartless desire, anyway.

Finally, the rule of restraint may have something to do with our intuitive fascination with complex mental states. Characters who are aware that other people are trying to read their body language, and hence attempt to control their body language to influence those people's perceptions of their mental states, may come across as more interesting than characters who simply let it all out. Restraint calls for a third-level embedment of mental states (as in, "I don't *want* her to *know* what I am *feeling*"), and, as I have suggested elsewhere, we may find particularly enjoyable cultural representations that cultivate this level of "sociocognitive complexity."[19]

Concealing Anger

Remember the two passages from *Persuasion* quoted above, one in which the author *tells* us what a character (Anne Elliott) feels, using the prerogative of omniscient narration, and another in which she makes a character's body *show* his true feelings, forcing him (Frederick Wentworth) into a state of embodied transparency? Here is another such juxtaposition of telling and making transparent, now from Austen's *Pride and Prejudice* (1813).

Mr. Darcy tells Elizabeth Bennet that he loves her and asks her to marry him. His proposal is not accepted kindly, and as the conversation goes on, both protagonists get angry. Note, however, that while we merely *hear* of Elizabeth's anger from the omniscient narrator, we actually *see* Mr. Darcy's anger and his attempts to conceal it.

Here is Elizabeth listening to Mr. Darcy's confession that he "struggled . . . in vain" to repress his love for her, a confession that may have been meant to flatter her but that she finds insulting:

> In spite of her deeply-rooted dislike, she could not be insensible to the compliment of such a man's affections, and though her intentions did not vary for an instant, she was first sorry for the pain he was [about] to receive; till, roused to resentment by his subsequent language, she lost all compassion in anger. She tried, however, to compose herself to answer him with patience, when he should have done.[20]

Elizabeth does not fully succeed in composing herself. Toward the end of Mr. Darcy's speech, "the colour [rises] into her cheeks" (fig. 2). But heightened color is not by itself a sign of anger. It can be interpreted as indicative of a variety of mental states, some even flattering to the suitor. In contrast, when, after hearing Elizabeth's response to his proposal, Mr. Darcy gets angry, his body provides direct and unequivocal access to his feelings (fig. 3):

> Mr. Darcy, who was leaning against the mantelpiece with his eyes fixed on her face, seemed to catch her words with no less resentment than surprise. His complexion became pale with anger, and the *disturbance of his mind was visible in every feature. He was struggling for the appearance of composure,* and would not open his lips, till he believed himself to have attained it. (129, emphasis added)

Observe our three rules at work in this scene (and as you do, apply them also to the passages from *Persuasion* discussed earlier). First, note the contrasts that go into constructing Mr. Darcy's transparency. Not only is he now more readable than Elizabeth, but he is also more readable than himself earlier in the novel and a moment before. In the first sentence of the last quoted passage, he is described as *seeming* to catch her

FIGURE 2. Jennifer Ehle as Elizabeth Bennet in the first proposal scene from the 1996 BBC production of *Pride and Prejudice*.

words with no less resentment than surprise. That is, there is still some possibility of misinterpreting his body language at that point: he *seems* to be resentful and surprised, but he might not actually be so. Then the next sentence ("His complexion became pale with anger . . .") leaves no doubt that his body reflects his mind fully and faithfully.

After a period of inner struggle (observe the rule of restraint in action!)—which leaves him transparent still, for the struggle is visible to Elizabeth—Darcy attains, or believes he has attained, "the appearance of composure." This quick sequence of contrasts, this change from seeming to being and then back to seeming—he *seems* resentful; he *is* angry; he *is* struggling; he *seems* composed—creates the impression that the moment of perfect access cannot last long, thus sharpening our appreciation for the vision of a body caught in spontaneous emotion.

True, Austen emphasizes that time slows down for Elizabeth as she watches Darcy's internal struggle—"the pause was to Elizabeth's feelings dreadful" (130)—but we know that this pause could not really have

FIGURE 3. Colin Firth as Mr. Darcy in the first proposal scene from the 1996 BBC production of *Pride and Prejudice*.

lasted very long. Brevity makes it ethically defensible, too: we don't want to think that Elizabeth actually enjoys watching Mr. Darcy at his most transparent.[21]

Anger, incidentally, is a particularly interesting emotion among those that a character may struggle to suppress. Social psychologist Larissa Z. Tiedens has demonstrated that anger appears to be "one of the few emotions for which suppression can actually foster affiliation—people like someone more if he or she restrains from displaying anger."[22] Fiction writers have always anticipated psychologists in their intuitive grasp of interpersonal dynamics. Here Austen does it again. We like Mr. Darcy; we have liked him for the last two hundred years. But now, with the advance of studies in cognitive and social psychology, we can see in a new light the small details that contribute to making him an appealing character, in spite of all his pride and snobbery.

This is not to say that concealing anger works like a charm in all fictional contexts. We are not automatically enamored of every character struggling to suppress his or her wrath. Rather, a moment of such suppression can increase our liking for a character when used strategically in conjunction with other rhetorical techniques.

Concealing Disappointment

Here is an example of embodied transparency from a more recent novel, Helen Fielding's *Bridget Jones: The Edge of Reason* (1999). Fielding's treatment of transparency is particularly interesting because, written from the first-person point of view and obsessed with the issue of gender and communication, her novel provides an apparently exhaustive report of Bridget's feelings and those of women surrounding her. (Men's minds remain strategically obscured—in the tradition of Austen, whose *Pride and Prejudice* and *Persuasion* inspired the *Bridget Jones's Diary* and *Bridget Jones: The Edge of Reason*.) There seems to be no need for special moments of embodied transparency.

But even here such moments are presented as rare and valuable flashes of insight. For example, there is the scene at a ski resort in which Bridget is talking to her boyfriend, Mark Darcy, and an attractive woman, Rebecca, who is trying to steal Mark from Bridget. Rebecca invites Mark and Bridget to a skiing party, where she would have more opportunities to flirt with Mark, especially if Bridget, a poor skier, could be separated from him:

> "Oh, it's so exhilarating," said Rebecca, putting her goggles on her head and laughing into Mark's face. "Listen, do you both want to have supper with us tonight? We are going to have a fondue up the mountain, then a torchlight ski down—oh sorry, Bridget, but you could come down in the cable car."
>
> "No," said Mark abruptly, "I missed Valentine's Day so I'm taking Bridget for a Valentine's dinner."
>
> The good thing about Rebecca is there is always a split second when she gives herself away by looking really pissed-off.
>
> "Okey-dokey, whatever, have a fun time," she said, flashed the toothpaste advert smile, then put her goggles on and skied off with a flourish towards the town.[23]

The rule of contrasts and the rule of transience are prominent here, more so than the rule of restraint. First, Rebecca's involuntary giving "herself away by looking really pissed-off" is contrasted with her fake spontaneity one moment earlier (when she appears unable to contain her good spirits buoyed by skiing) and her fake friendliness right after (when she smiles broadly to show that she does not mind Mark's rejection). Her bodily display of feelings is also contrasted with Mark's opacity. When Mark says "no," it comes across as "abrupt," which means that nothing in his body language has prepared the two women for what he was about to say.

Second, Fielding has Bridget actively draw our attention to the transience of this revelatory moment: Rebecca looks pissed off only for a "split second," so one is lucky to catch it. Everything happens too fast for anybody to register Rebecca's struggle to hide her disappointment. Hence we get no explicit description of such a struggle. Do we still read it into the scene? Perhaps it is there, though not described explicitly, hiding between Rebecca's looking pissed off and flashing the toothpaste advert smile?

Magical Exceptions

Now think of narratives whose genre conventions exempt them from the rule of transience, such as myths and fairy tales or modern-day stories with elements of magic. For instance, in the movie *What Women Want* (2000) Mel Gibson's character, Nick Marshall, falls into a bathtub with a hairdryer and is jolted by electricity, which makes it possible for him to hear the innermost thoughts of women around him. This fantastic premise makes the issue of transience moot: women are transparent to Nick *all the time.*

What electricity does to Nick, fairies do to the knight from the medieval French fabliau "Le chevalier qui fist parler les cons" ("The Knight Who Made Cunts Speak"). They give the poor but "gallant" protagonist a marvelous gift: he can literally force women's body parts to talk against their owners' wills. As one of the fairies explains:

> Sir knight, my gift's no small one:
> wherever you go, west or east,

you shall not find a maid or a beast,
so she have two eyes, whose cunt can refrain
from answering you if you but deign
to speak to it.

Another fairy continues:

Sir knight, to this second gift I add,
as is just and right, that if the cunt
be blocked or stoppered up in front
and cannot answer you straightway,
the arsehole will, without delay,
speak for it, if you give leave,
no matter whom it hurt or grieve.[24]

The knight soon finds himself a guest in the castle of a count and countess where he has ample opportunity to display his new magical powers to the mortification of his hostess and her female servants. As the scholar of French literature Evelyn Birge Vitz observes, this "fabliau enacts the desire (of men) to hear the lower, interior parts of the female body speak up—and for these parts, unlike women's mouths, to tell the truth."[25] In other words the same intense interest in correlating body language with "true" states of mind drives depictions of embodied transparency in a novel by Jane Austen and in a bawdy medieval tale. The fantastic premise of the latter, however, makes transience less of an issue: to the embarrassment of the female protagonists their lower body parts can, in principle, hold forth for *any* length of time.

I expect to find the same disregard for the rule of transience in tales featuring supernatural agents and events in a variety of historical periods and national literatures. The underlying assumption here is that authors always look for new ways to bring about embodied transparency and that when they work with genres that allow the bending of reality, they use that license with vengeance to make bodies transparent. If we are reading, watching, or listening to a story in which magic allows the protagonist to know what others think, we are with him or her all the way—we want as much of it as we can get.

At least for a while, that is. For, ethics may start whimpering in the background, and the author may feel compelled to tax the protagonist with the sin of pride, which then necessitates punishment and rescinding of the magic powers.

Who Gets to See Transparent Bodies?

It may seem, based on my examples from Austen, Helen Fielding, *What Women Want*, and "The Knight Who Made Cunts Speak," that embodied transparency always finds an appreciative spectator within the story itself. When Mr. Darcy looks angry, Elizabeth is there to observe him and interpret his body language. When Rebecca looks pissed off, Bridget is there to notice it. When Helen Hunt's character is brainstorming an advertising slogan, Mel Gibson's character is there to eavesdrop on her thoughts and steal them for his own professional advancement. And the knight is there when the countess's . . . you get the picture. But the situation is really more complex. Many fictional characters remain oblivious to others' momentary transparency because they do not have enough context to interpret it.

Thus in Henry Fielding's *Tom Jones* (1749) there is a scene in which the virtuous Mr. Allworthy, the adopted father of the title character, is talking about love as "the only foundation of happiness in a married state." Present at this heartfelt "sermon" is one Doctor Blifil, the brother of the man who is about to marry Mr. Allworthy's sister, Bridget. Doctor Blifil knows that his brother is marrying not for love but in hopes of inheriting Mr. Allworthy's estate, so it costs him some effort to refrain from sneering as he listens to the good man's idealistic exhortations. As Fielding puts it, Doctor Blifil has to take "some pains to preserve now and then a small discomposure of his muscles."[26]

We infer that Allworthy remains oblivious to these half-smothered facial contortions.[27] Or, perhaps, he does see them but perceives them as mere facial ticks. In any case they do not have for him the same meaning they have for us. We look at Doctor Blifil, and we see what Mr. Allworthy does not see: a body that desperately does not want to be read and thus *is* readable in this desire not to be read. In that moment Doctor Blifil is transparent, but we are the only appreciative audience for his transparency.

We know what the Doctor's writhing mug entails because Fielding wants us to know. In contrast consider another occasion when a character is going through an orgy of grimacing to conceal his feelings, but Fielding does not want us to know what it means, at least not until we come to the end of the book.

Bridget Allworthy does marry the venal Blifil. Eventually, both die. Their son, the young Blifil, plots to make Tom Jones look bad in Mr. Allworthy's eyes. He finally succeeds, and Tom is kicked off Mr. Allworthy's estate. During his subsequent journey to London Tom comes across an old Gloucester acquaintance, Mr. Dowling, the lawyer. The two men sit down to a bottle of wine. At some point the conversation turns to Blifil, and when Dowling learns of Tom's poor opinion of that young man, he observes that "it is a pity such a person should inherit the great estate of your uncle Allworthy." Tom's reply—"Alas, sir, you do me an honour to which I have no title. . . . I assure you, sir, I am no relation of Mr. Allworthy"—makes it clear that he doesn't know that on her deathbed Bridget Blifil confessed that Tom is her illegitimate son and hence Allworthy's nephew.

As Dowling listens to Tom's earnest professions that he has never thought himself entitled to any part of Allworthy's estate, he realizes that the young man in front of him has been cheated out of his family and fortune. For Dowling was there with Bridget as she lay dying and wrote to Allworthy about her confession. If Tom is still ignorant about his origins, it means that Allworthy himself or someone in his household (perhaps the young Blifil?) intentionally suppressed the information.

Surprised as he is by what he hears, Dowling is not about to disclose the truth now. Instead, he tries "to hide" his feelings from Tom "by winking, nodding, sneering, and grinning" (576). We see his grimacing, but we don't know what it means. We do not learn about Tom's parentage for another three hundred pages. Only after we finish the book can we go back to that scene in the tavern and realize that Dowling's body language was making him transparent—that he was perfectly readable in his desperate desire not to be readable—but there was nobody there with enough knowledge of the situation to appreciate his struggle to conceal his feelings.

We are entering interesting territory with this last example of transparency that characters don't see at all and readers can appreciate only when they reread the novel. We can call it "protocinematic" because this is what movies often do. They make us observe a character's body language—not perceived by other characters—but we don't learn until later what it meant. Or we call it protodetective—as in detective stories. For there, too, once the secret is finally revealed, we have to go back and rethink the clues, including people's body language. (And, indeed, because of various clues scattered throughout *Tom Jones,* Fielding's novel has been called an early example of a detective narrative.)

Conclusion: What's in It for Us?

In the next chapter I turn to literary examples of embodied transparency that range from ethically iffy to disturbing. Before we move on, however, here is a practical question to consider: what's the payoff of foisting this new concept onto readers, especially those weary of the proliferation of jargon in cultural criticism?

The concept of embodied transparency is useful because it allows us to recognize a common pattern that cuts across genres, historical periods, and national representational traditions, a pattern rooted in our imperfect and powerful adaptations for mind-reading. After all, we wouldn't usually consider side by side a medieval tale about a knight whose special talent cannot be described without using four-letter words, a movie premised on Mel Gibson's character falling into a bathtub, a reality show in which a participant is put in an embarrassing situation and filmed in close-up as she becomes aware of just *how* embarrassing the situation is, and a Jane Austen novel in which one protagonist is closely observed by another as he glances up and smiles almost imperceptibly in response to someone's remark.

Recognizing these diverse narratives as creating contexts for moments of embodied transparency does not detract from their historical uniqueness or their complex involvement with their respective genres—and why should it? It does, however, allow us to recognize ourselves as always trying to imagine what things would be like were bodies perfectly readable—laughing at some of our imaginings, sighing wistfully at others.

And yet—here is something to keep in mind in the middle of this talk of embodied transparency as a recurrent pattern: it *is* relatively rare in prose fiction, especially compared to other techniques used to give us direct access to characters' feelings. Cognitive narratologist Alan Palmer points out that one "of the pleasures of reading novels is the enjoyment of being told what a variety of fictional people are thinking. . . . This is a relief from the business of real life, much of which requires the ability to decode accurately the behavior of others."[28] Because it's important that we don't start seeing embodied transparency at the drop of a hat, I will close with more examples of writers telling us what "fictional people are thinking" without recourse to embodied transparency (even when characters' bodies are described in some detail):

"Levin was insufferably bored with the ladies that evening"—this is Lev Tolstoy in *Anna Karenina*, using third-person omniscient narration.[29]

"Conscience that had slept so long, begun to awake, and I began to reproach my self with my past life, in which I had so evidently by uncommon wickedness, provok'd the justice of God to lay me under uncommon strokes, and to deal with me in so vindictive a manner"—this is Daniel Defoe in *Robinson Crusoe* using the technique of first-person narrator.[30]

"The hair was curled, and the maid sent away, and Emma sat down to think and be miserable. It was a wretched business indeed. Such an overflow of everything she had been wishing for. Such a development of everything most unwelcome! Such a blow for Harriet!"—this is Austen in *Emma*, using the technique of free indirect discourse.[31]

"That night in the mess after the spaghetti course, which every one ate very quickly and seriously, lifting the spaghetti on the fork until the loose strands hung clear then lowering it into the mouth, or else using a continuous lift and sucking into the mouth, helping ourselves to wine from the grass-covered gallon flask; it swung in a metal cradle and you pulled the neck of the flask down with the forefinger and the wine, clear red, tannic and lovely, poured out into the glass held with the same hand; after this course, the captain commenced picking on the priest"—this is Hemingway in *A Farewell to Arms*, using his trademark technique of forcing readers to intuit mental states behind his characters' nondescript behavior.[32]

A novel can thus feature third- or first-person narration, or free indirect discourse, or it can force us to guess the mental states behind ambiguous body language, or it can rely on any combination of these techniques. On top of that, it may (or may not!) include scenes in which the character's body speaks her mind directly, and these are the moments that I am after.

THREE

‖‖‖

{ In which readers encounter a third writer named Fielding / a rich man plays with a poor man's feelings while a beautiful girl watches / an old man plays with a young man's feelings while a beautiful girl has no clue / and the protagonist of *Fight Club* shows that he cares. }

Sadistic Benefactors

Whom best I love, I cross; to make my gift,
The more delay'd, delighted.
       —Shakespeare, *Cymbeline* 5.4.101–2

Some fictional characters get a glimpse of other characters' true feelings; others don't notice a thing; and some may even get a chance to think back and realize what this or that look or gesture truly meant.

There is also another category of characters—those who are not content with merely glimpsing other people's feelings. Instead, they want to script such moments of transparency themselves. That is, they want to *force* others into revealing their feelings through body language.

This last emotional terrain is often explored by horror stories and psychological thrillers. As literary critic Walter Benn Michaels puts it in his discussion of *American Psycho*, "You can be confident that the girl screaming when you shoot her with a nail gun is not performing (in the sense of faking) her pain."[1] To avoid or soften the charge of sadism leveled against characters who instigate embodied transparency in others, their actions can be shown to be driven by revenge (as in *The Count of Monte Cristo*) or, paradoxically, by affection or desire to do good.

The ethics of the latter situation are extremely ambiguous. An undertow of emotional sadism runs through them. Yet such characters don't engage in what can be characterized as straightforward torture. In fact, they may sincerely believe that their actions will ultimately benefit the people whom they are forcing into transparency. To reflect this ambiguity, I call such characters sadistic benefactors.[2]

One sadistic benefactor makes his appearance in a work by yet another Fielding: Sarah, Henry's sister. In her novel *The History of Ophelia* (1760) a rich man sends a poor man on an emotional roller coaster to enjoy the spectacle of his feelings. Yet he acts on the benevolent principle of Jupiter from Shakespeare's *Cymbeline*: he "crosses," that is, inflicts pain on, those whom he loves "best" in order to intensify their subsequent joy. (I bring up Jupiter on purpose. When it comes to deities, their supernatural powers create endless opportunities for forcing mortals into embodied transparency. Were I to write about fictions of transparency in the ancient world, gods would figure in my account most prominently.)

Here is how Fielding builds up to that eventual joy. The novel's protagonist, Lord Dorchester, comes across a starving half-pay soldier, Captain Traverse, and decides to help him. Through his connections at court, Dorchester secretly procures Traverse a choice of two jobs. He begins, however, by telling the captain only of the first job, one that Dorchester knows Traverse will not be able to take because of family circumstances. The poor captain, unwilling to appear ungrateful, receives "this News with as much Gratitude as if it had been the very Thing he wished" and turns it down politely. Lord Dorchester then expresses his disappointment in such guilt-inducing terms as to drive the captain to break down in tears when he thinks nobody is watching. (In fact, several people are watching, including Lord Dorchester's love interest, Ophelia, the young girl who is telling the story.)[3]

Not yet content with this show of emotion, Lord Dorchester then reveals the captain's family waiting in the next room and urges him again to take the first job. In response Traverse "faint[s] away instantly," terrifying his wife and making the onlookers fear for his life. When he comes to, Lord Dorchester augments "the general Joy" that his recovery occasions by telling him of the second job, one that is completely acceptable and will save the whole family from starvation. The joy now increases "to a great Degree of Extacy," rising to a "Height that must have been painful." The captain and his wife look on the "Lord with Adoration, and [give] way to Raptures that would have forced a Heart the most insensible to the Sensations of others, to partake of theirs" (1:254–55).

We can call Dorchester's behavior sadistic and think of him as a mini-Jupiter of that claustrophobic world, but there is no way of telling if Fielding herself viewed him thus.[4] All we can safely assume about her thoughts on the subject is that, as many writers before and after her, she was intuitively interested in figuring out how to put characters in situations in which their bodies reveal their minds. She drew on available cultural contexts of her day: a fascination with private philanthropy, an obsession with social class, and the conventions of contemporary sentimental discourse (in which men may cry and faint and express their emotions relatively freely). Admittedly, she may have pushed these conventions a bit too far in having Dorchester mastermind the touching scene instead of letting it happen by chance.

Two years later, another eighteenth-century writer profoundly invested in sentimentalism brought forth an even more manipulative sadistic benefactor. Jean-Jacques, the narrator of Rousseau's *Emile* (1762), is a strange case. He is a hybrid between fictional character and Rousseau's own self; other people in the novel are figments of Jean-Jacques's imagination—he makes them up as he moves along to illustrate his philosophical and pedagogical points. He tells us, too, that he makes them up, but we soon forget this and, courtesy of our theory of mind, begin treating them not as abstract entities used for object lessons but as regular fictional characters (that is, as independent agents capable of a rich array of thoughts and feelings).

One such character is a young man named Emile, who has been under Jean-Jacques's tutelage from early childhood. Toward the end of the novel Emile meets the woman of his dreams, Sophie. Now she is all he can think about. Jean-Jacques wholeheartedly approves of his choice, for it was he who brought the young people together (indeed, he invented Sophie just for this occasion and told us so). Still, he also believes that his pupil has yet much to learn about himself and the world. Jean-Jacques thus needs to convince Emile that he has to leave Sophie for two years in order to complete his education as a man and as a citizen. *Then* he can come back to her ready to assume the duties of the head of the family. This is a difficult case to argue, and Jean-Jacques decides that, in order to win, he first has to put Emile in a state of emotional turmoil seemingly unrelated to the topic at hand. Here is how he goes about it:

> One morning, when they have not seen each other for two days, I enter Emile's room with a letter in my hand; staring fixedly at him, I say, "What would you do if you were informed that Sophie is dead?" He lets out a great cry, gets up, striking his hands together, and looks wild-eyed at me without saying a single word. "Respond then," I continue with the same tranquility. Then, irritated by my coolness, Emile approaches, his eyes inflamed with anger, and stops in an almost threatening posture: "What would I do . . . I don't know. But what I do know is that I would never again in my life see the man who had informed me." "Reassure yourself," I respond smiling. "She is alive. She is well. She thinks of you, and we are expected this evening. But let us go and take a stroll, and we will chat."
>
> The passion with which he is preoccupied no longer permits him to give himself to purely reasoned conversations as he had before. I have to interest him by this very passion to make him attentive to my lessons. This is what I have done by this terrible preamble. I am now quite sure that he will listen to me.[5]

Readers familiar with the story of Jean-Jacques's relationship with Emile won't be surprised by his approach. Jean-Jacques has raised Emile in such a way that he can read "in his face all the movements of his soul" (226). This is to say, he has been manipulating the boy since their first days together. Still, the above passage stands out in its emotional violence. The tutor's "cool" intimations that Sophie might be dead come closer to sadism than anything else he has ever done to his impressionable charge.

We know why he is doing it. Because of his passion for Sophie, Emile has recently been less available to his tutor. Jean-Jacques needs to restore Emile's absolute transparency. Forcing him onto an emotional roller coaster—she is dead; no, she is alive and thinking about you—achieves that goal. Note how the text testifies to the tutor's regained ability to read Emile's mind perfectly: Jean-Jacques thinks that Emile is "irritated by [his] coolness," angry with him, and "almost" ready to attack him. When, shortly after, the young man speaks, his words prove that Jean-Jacques has been reading his body language correctly: Emile says that he would remain forever hostile toward a man who informs him of Sophie's death.

Jean-Jacques's treatment of Emile would be unforgivable had he not acted in the young man's best interests. It makes Emile transparent and pliable, open to Jean-Jacques's subsequent arguments about the importance of parting with Sophie for two years. The temporary anguish into which Emile is plunged is thus a precondition for his future happiness. Moreover, by this point in the story we have no doubt that Jean-Jacques loves Emile more than anything in the world—albeit because he does not have anybody else in the world. This overarching narrative of love is called forth to excuse and redeem the act of forcing emotional transparency onto the young man. The benefit vastly outweighs the cost, and Jean-Jacques remains beloved by Emile, Sophie, and (later) their children.

Enter present-day sadistic benefactor Tyler Durden, from Chuck Palahniuk's *Fight Club* (1996). Tyler holds a gun to the head of a man he's just met (who turns out to be a twenty-three-year-old college dropout) and extracts from him the promise that he will go back to school to finish his degree in veterinary medicine. Tyler wants to impress on "Raymond Hessel" (the name he reads off the man's driver's license) an important life lesson: death can strike any minute, so study hard and follow your dreams.

But before Tyler gets to the salutary follow-your-dreams part, he uses every grisly cliché to convince Raymond of his imminent and terrible demise. He explains to the crying man how he will "cool" down, passing from a "person" to an "object" and how his "Mom and Dad would have to call old doctor whoever and get [their son's] dental records because there wouldn't be much left of [his] face." The scene is structured so that Tyler seems to merely report the images arising in his victim's mind. Raymond is forced into embodied transparency: half-paralyzed with fear, crying harder and harder, following meekly Tyler's orders, thinking of things that Tyler tells him to think of (fig. 4).

When Tyler finally lets Raymond go, his thoughts stay on him as he muses with great satisfaction: "Raymond fucking Hessel, your dinner is going to taste better than any meal you've ever eaten, and tomorrow will be the most beautiful day of your entire life."[6] Like Jupiter and Lord Dorchester, Tyler apparently believes that to intensify somebody's happiness you must first make them truly miserable. More important, like

FIGURE 4. *Left to right:* Joon B. Kim, Brad Pitt, and Edward Norton in the "Raymond Hessel" scene from David Fincher's *Fight Club.*

Lord Dorchester, Tyler treasures every moment of transparency he can wring out of his victim. He enjoys knowing exactly what Raymond is thinking now and what he will be thinking tomorrow.

Or does he?

In *Fight Club* Palahniuk uses the literary convention of unreliable narration; that is, we can't trust the narrator's account of events.[7] Bearing this in mind, look again at the scene in which Tyler reads his victim's body as an open book: "You [are] going to cool, the amazing miracle of death. One minute, you're a person, the next minute you're an object, and Mom and Dad would have to call old doctor whoever and get your dental records because there wouldn't be much left of your face, and Mom and Dad, they'd always expected so much more from you and, no, life wasn't fair and now it was come to this" (153).

If we return to this scene after we have finished the novel and found out that Tyler is an unreliable narrator, we may start seeing problems with this assured interpretation of Raymond's body language. We may notice, for example, that Tyler's account of Raymond's thoughts draws on conventional images provided by crime dramas. We have a visual repertoire of scenes associated with violent deaths, for example: cut to the bereaved family; cut to the corpse in the morgue; cut to the dentist, or some other doctor, confirming the victim's identity; and so forth. This stale, depersonalized repertoire is what Tyler dips into for his "report" from Raymond's head.

Similarly, when Tyler confidently foretells what Raymond will be thinking even after he is out of Tyler's clutches ("tomorrow will be the most beautiful day of your entire life"), I see his point. I can certainly imagine how tomorrow Raymond might feel almost unbearably happy to be alive and thankful for every crumb that passes his lips and for every leaf that he sees trembling in the wind. I can also imagine Raymond racing to school, profoundly grateful for the opportunity to "work [his] ass off" (154) on various difficult subjects, just as Tyler told him he should. Finally, I can imagine Raymond eventually becoming a successful veterinarian, loved by his family, respected by his neighbors, and remembering now and then with wonder and gratitude that fateful moment when a stranger with a gun forced him to turn his life around and make the most of it.

But then I can also imagine Raymond falling into a profound depression soon after his encounter with Tyler, thinking obsessively that his life depends on the whim of some jerk with a gun, and killing himself one day after school.

Or he may get a gun of his own and hunt down Tyler.[8]

In other words Raymond's body remains transparent and his mind accessible as long as we consider this scene in isolation from the tradition of unreliable narration. For within this tradition, when a first-person narrator reports another character's thoughts, he is almost immediately suspect, and his reporting must be scrutinized for signs of inconsistency, vested interests, or madness. So Palahniuk's readers believe that Tyler really knows what Raymond is thinking only as long as they are not aware that Tyler is an unreliable narrator. Once they are aware of it, they have the option—which, of course, they may not choose—of assuming that they have learned little about Raymond's actual feelings on the occasion.

Incidentally, even if we do not consider this episode in relation to the convention of unreliable narration, something else in it alerts us to the likely gap between Tyler's assured interpretation of Raymond's body and Raymond's actual mental state. That something else is the violation of the rule of transience, the second rule for constructing scenes of embodied transparency. Tyler keeps reading Raymond's mind for three straight pages. This implies transparency enduring far beyond what we've seen in our other examples, in which transparency lasted for a split second or several seconds.

Of course, Tyler may believe, egomaniac that he is, that he can keep Raymond transparent for as long as he pleases. The longer they are conversing, however, the more open we are to the possibility that after a while Raymond intuits something about the twisted psychology of his captor and starts performing his fear and despair at the top of his lungs. This is just a speculation. I don't have any direct textual evidence that Raymond is performing his feelings. It just seems to me that when embodied transparency is induced, observed, and reported by an unreliable narrator, readers may begin to wonder. Can't transparency—if it seems to go on and on—morph into performance without the unreliable narrator's noticing it?

The literary convention of unreliable narration thus has an interesting relationship with embodied transparency. On the one hand, this convention offers writers more opportunities for putting characters into situations in which their bodies betray their feelings. On the other hand, toward the end of the story readers often realize that they cannot trust any accounts of embodied transparency if the voice behind those accounts has been that of an unreliable narrator.[9]

This, in turn, raises a question about the relationship between mind reading and power. Of the three sadistic benefactors discussed in this chapter—four, if we count Jupiter—Tyler is the only one who has no real power over his victim outside of the immediate (i.e., gun-wielding) context. Jupiter can torture mortals because he is a god. Lord Dorchester can torture Captain Traverse because, for a poor soldier in eighteenth-century England, a lord is the closest thing to a god. Jean-Jacques can torture Emile because he is the only adult in charge of this otherwise effectively parentless child. (Also, speaking of gods, Jean-Jacques created Emile and told us so.) Tyler has nothing going for him, except the gun, the empty street, and the stories that he tells himself about his power to read people's minds and change their lives.

The glaring mind-reading asymmetry implied by sadistic benefaction (e.g., Lord Dorchester manipulates Captain Traverse's mind while Captain Traverse has no access to Lord Dorchester's mind) is thus always a reflection of an existing power asymmetry: gods vs. mortals, rich vs. poor, adults vs. children. Access to minds means power; the effective manipulation of minds constitutes abuse of this power.[10] Unreliable narrators such as Tyler, who have no objective claim to any kind of power (godhead, riches, or parenthood), go directly for its characteristic manifestation: the ability to control other minds. To the extent to which we believe that they succeed, we underappreciate their unreliability; that is, we invest them with more power than they actually have.

FOUR

In which Clarissa is fooled while Evelina watches a fool / Lev Tolstoy, Ernst Lubitsch, and Alfred Hitchcock walk into a hippodrome / and readers are made to think first of Cary Grant and then of Colin Firth.

Theaters, Hippodromes, and Other Mousetraps

Why Going to the Theater Is Good for a Story

Why do we go to the theater? A theory-of-mind hardliner such as myself, determined to see everything through the lens of mind reading, would say we go to the theater to give our greedy theory of mind a very particular and rich treat. We watch actors' facial expressions and body language; we correlate them with given social contexts; and we follow sequences of emotions displayed by one actor over time and ranges of emotion displayed by different actors at the same time. That is, we go to the theater to *feel*—to experience a rich gamut of emotions while surrounded by people who are going through similar sensations—but this complex emotional experience is inextricably bound with our reading of the characters' mental states.

Why do fictional characters go to the theater? They do it because the author has to develop the story, and sending protagonists to the theater propels the plot forward. Theater is a place where they can run into other people, such as lovers or enemies; where they can hear important gossip, get kidnapped, become aware of their changed social status, and leave or receive an indelible first impression.

It is also a place where they can catch a glimpse of unguarded body language and thus learn something important about other characters' feelings.

You may notice as you read various fictional accounts of the theatergoing experience, that sometimes there is a spectator in the audience who does not watch the stage or only watches it with one eye while keeping the other on fellow spectators and their involuntary reactions to the play. There is Hamlet closely observing his uncle during the loaded stage reenactment of "The Murder of Gonzago" and interpreting the king's spontaneous body language as proof of his guilt. There is Clarissa, from Samuel Richardson's eponymous novel (1747–48), watching a play and feeling "greatly moved by it" yet also, out of the corner of her eye, checking whether or not the man she is with, Robert Lovelace, is "sensibly touched with some of the most affecting scenes."[1] There is Katie Carr, a protagonist of Nick Hornby's novel *How to Be Good* (2001), loving "every second of the play" yet spending "almost as much time" observing the complex emotions written on her husband's face as he follows the performance.[2]

This practice of surreptitious observation is not accidental. It adds up to an important narrative convention: bringing protagonists to the theater opens up yet another possibility for embodied transparency. This convention depends on a very particular cultural assumption about theatergoing: people go to the theater to watch the characters onstage and interpret their behavior, which means that for the duration of the performance they let down their guard because nobody is watching *them* and interpreting *their* behavior. That is, in a social setting that clearly marks off some people as performers and others as spectators, and in which the performers' success is judged by the spectators' attention (the best performances keep their audiences spellbound), spectators relax and let their bodies show their emotions.

This means, of course, that if somebody else in the audience happens to ignore the actors and focuses instead on fellow spectators, that person can learn quite a bit about another's true feelings.

At least this is how it works for Hamlet, Clarissa, and Katie Carr. Once more, we have to remember the difference between embodied transparency in real life and in fiction. In real life I don't think I've ever tried to learn anything about another person's state of mind by watching them as they watch a play. First of all, it's dark. And even when it's not dark (as in smaller, experimental theaters), I can only see part of a person's face, and I am more interested in what's happening onstage, anyway.

Also, the negative correlation between social complexity and transparency is still valid. Let's say I do observe some spontaneous body language of the person I'm with. The more interesting, that is, the more socially complex, narrative I come up with to interpret it, the more likely I am to be wrong. OK—he dabs a tear during that affecting scene, so this means . . . what? That he feels bad about making me feel lonely two weeks ago when we talked about X—and this is because the heroine's predicament reminds me in some complicated ways of my own, so I assume that he, too, is thinking about it this way? Right . . .

Not so with the fictional theatergoers. They can read complex mental states into the faces of other spectators with perfect fluency. Here is Katie Carr interpreting her husband's spontaneous reaction to a play with an almost disconcerting assurance. David has always hated theater, but he is determined to enjoy it this time. Katie finds his inner struggle, plainly written on his face, as gripping and informative as what is taking place onstage:

> I love every second of the play. I drink it, like someone with dehydration might drink a glass of iced water. I love being made to think about something else other than my work and my marriage, and I love its wit and its seriousness, and I vow for the millionth time to nourish myself in this way on a more regular basis. . . . I spend almost as much time trying to snatch glimpses of David's profile as I do watching the stage, though. Something weird has happened, definitely, because the struggle to enjoy the evening is written on David's face: a war is taking place there, around the eyes and lips and the forehead. The old David wants to frown and scowl and make faces to indicate his contempt for everything; the new one is clearly trying to learn how to enjoy himself in a place of entertainment, watching a new and brilliant piece of work from one of the world's leading playwrights. (69)

Readers of *How to Be Good* may remember that David has just undergone a personality change and that Katie finds the thought processes of the "new" David so puzzling that "he is beginning to give [her] the creeps" (73). So the personal dynamic of this scene contrasts sharply— and, for Katie, pleasurably—with the current goings-on in the Carr household. Whereas at home Katie feels that she doesn't "really know" David, at the theater she can read him as an open book.

Remember the last time something like this happened to you? Me neither.

Performance Creeps Back In

To recap: when a writer wants to put a character into a situation in which his or her mental state is radically legible to an interested party (especially if that party is generally unsure what the other is thinking), bringing them both to the theater is one viable narrative strategy. Theater is good for a story because it can create an immediate opening for embodied transparency.

This recipe for privileged mind reading is not foolproof, however. In fact, no such recipe can remain foolproof for long because—remember?—as the best source of information about the mind, the body is also the most suspect source of information about the mind. Whenever a cultural setting becomes a recognizable context for fictional embodied transparency, it is immediately ready for subversion. Precisely because it is now known as a space where characters' bodies leak their feelings, it can be used for a more devious performance of these feelings.

Theater in particular has long been vulnerable to this kind of double billing. The same novel can include one scene in which a character observes another character at the theater and learns something crucial about him based on his spontaneous body language; and another scene in which a character goes to the theater knowing that his body will be scrutinized for spontaneous emotions and carefully stages his display of emotions to manipulate the naive observer.

In *How to Be Good* Hornby chooses to keep the matter simple and to portray the theater as a reliable setting for embodied transparency. So, obviously, this option is still available to a writer. See, however, what happens in another novel, written two and a half centuries ago, which features not one but two protagonists uncertain about each other's thoughts and feelings.

Richardson's *Clarissa* relies on the sentimental cliché of theater as catalyst for spontaneous displays of emotion and simultaneously subverts this cliché. At one point the novel's villain, Robert Lovelace, invites Clarissa Harlowe, an eighteen-year-old paragon of virtue, and his love interest, to see a tragedy. Lovelace is certain that however affected Clarissa might be by the play, she will also be watching him and judging his moral worth and true sentiments (of which she is not at all sure) by his reaction to what is happening onstage.

Lovelace is ready to put on a suitable performance of his exalted feelings. And not just that. He also takes along his accomplice, a prostitute named Polly, whom he passes off as an upright and sensitive young lady. Clarissa is thus to be impressed both by Lovelace and the company he keeps. Polly, however, needs some coaching, and Lovelace provides it. As he reports in a letter to his confidant, "I have directed [Polly] where to weep—and this not only to show her humanity (a weeping eye indicates a gentle heart), but to have a pretence to hide her face with her fan or handkerchief."[3]

The plan works beautifully. As Clarissa reports afterwards in a letter to *her* confidante: "I was at the play last night with Mr. Lovelace and Miss Horton [i.e., Polly]. It is, you know, a deep and most affecting tragedy in the reading. . . . You will not wonder that Miss Horton, as well as I, was greatly moved at the representation, when I tell you, and have some pleasure in telling you, that Mr. Lovelace himself was very sensibly touched with some of the most affecting scenes. I mention this in praise of the author's performance; for I take Mr. Lovelace to be one of the most hard-hearted men in the world" (640).

What Richardson appears to be saying here is that only naives, such as Clarissa, still believe that people stop performing their bodies once they direct their attention to the stage. Lovelace and Polly seem to know that the real performance—and the most exquisite fakery—only begins then.

Except that it is not that simple. Here and elsewhere in *Clarissa* Richardson manages to send up sentimentalist assumptions and to rely on them at the same time. True, Polly successfully fakes a kind heart and lofty sensibility, and so does Lovelace. But even as he cynically plots their "involuntary" behavior in the theater, Lovelace inadvertently admits that he actually believes that watching a play can and will reveal the true nature of a spectator. When he says that one reason he "directed [Polly] where to weep" was that she would have a "pretence to hide her face with her fan or handkerchief," this means that he expects that Polly will be naturally inclined to laugh during the "most affecting scenes" and thus will need a fan or handkerchief to cover her face lest her laughter betray her actual lack of kindness, good understanding, and virtue.

For, as Lovelace adds in the same breath, theater does carry the "heart . . . out of itself" (640), and spontaneous nonscripted bodily reactions to what happens onstage do reveal something about a person. Polly's laughter *will* show her true self unless she takes care to cover it up. In other words theater does work as a context for embodied transparency, but it is fragile and vulnerable to subversion (and Clarissa learns this to her cost).

What makes this particular theatergoing experience especially open to manipulation is that Clarissa constantly searches for deeper meaning in everything that Lovelace does. His every action *must* stand for something else, revealing some essential moral quality. So she can't simply go to the theater with Lovelace and enjoy the show: she has to observe and judge him all the time (fig. 5). It is almost inevitable, then, that knowing that she does this, Lovelace would use the occasion to pretend to embody the moral virtues that he wants her to think he possesses. Here and elsewhere in the novel, so much is at stake for Lovelace and Clarissa—so desperately does each want to figure out what the other is thinking and to bend the other's will to his or her own—that every opening for embodied transparency becomes instead an opportunity for a targeted performance.

Unruly Audiences

"For my part," said Mr. Lovel, "I confess I seldom listen to the players: one has so much to do, in looking about, and finding out one's acquaintance, that, really, one has no time to mind the stage. Pray,"—(most affectionately fixing his eyes upon a diamond-ring on his little finger) "pray—what was the play to-night?"

FIGURE 5. Clarissa (Saskia Wickham) observes Lovelace (Sean Bean) at the theater in the 1991 BBC series *Clarissa*.

> "Why, what the D——l,"—cried the Captain, "do you come to the play, without knowing what it is?"
>
> "O yes, Sir, yes, very frequently: I have no time to read play-bills; one merely comes to meet one's friends, and shew that one's alive."
>
> —Frances Burney, *Evelina*

Is it possible that in some cultures theater would *not* be perceived as a convincing context for embodied transparency in fiction? After all, my argument assumes a very particular type of spectators: those who sit quietly and follow the events on the stage with bated breath, losing themselves in these events to such a degree as to forget to control their body language. This picture may reflect (and idealize) our own cultural practices but not those of theatergoers in other historical periods.

Eighteenth-century novels such as Richardson's *Clarissa* or Frances Burney's *Evelina* (1778) represent good test cases for this question because eighteenth-century English spectators did not treat events onstage with the reverence that we do. That is, then as now, people certainly went to the theater to *feel,* to have the experience of the heart carried "out of itself." To a much greater degree than we do today, however, they also saw in theater an opportunity to socialize: to catch up with the latest gossip and to make new acquaintances. People attended the same play multiple times—not necessarily because they admired the acting but because theater was the place to go in the evening to see one's friends, and there weren't many other entertainment outlets available. Spectators thought little about talking to others during the performance, which meant that actors had to speak their lines over the constant din rising from the audience.

It matters, too, that today, once the play begins, the lights in the house go off, so we can't see each other, whereas back then, the house was fully lit the whole time, which meant that members of the audience could see each other and read each other's body language to their hearts' content. Presumably, this must have rendered them more self-conscious and less likely to "lose" themselves in the events onstage.

This is not to say that eighteenth-century theatergoers did not care about what happened on the stage. They did. Not paying *any* attention to acting was considered affectation. Mr. Lovel from *Evelina,* who claims to "seldom listen to the players," is a pretentious fop. (In contrast, the novel's sympathetic protagonist, a sensitive young girl named Evelina, is shown to lose herself in a particularly "slow and pathetic" aria at the Opera.)[4] However noisy and unruly eighteenth-century theater may seem by our standards, good performances did command the audience's attention, while *great* performances—for example, by Thomas Betterton, Barton Booth, David Garrick, or Sarah Siddons—were known to make the whole house hold its collective breath and remain completely quiet for long periods.

Henry Siddons, the son of Sarah Siddons and himself an actor and playwright, left us the following meditation on the range of behaviors exhibited by late eighteenth-century and early nineteenth-century theatergoers:

When a person sits at the theatre, after having seen a play acted
three or four times, his mind naturally becomes vacant and inactive.

If among the spectators he chances to recognize a youth, to whom the same is new, this object affords him, and many others, a more entertaining fund of observation than all that is going forward on the stage.

This novice of an auditor, carried away by the illusion, imitates all he sees, even to the actions of the players, though in a mode less decisive. Without knowing what is going to be said, he is serious, or contented, according to the tone which the performers happen to take. His eyes become a mirror, faithfully reflecting the varying gestures of the several personages concerned.

Ill humour, irony, anger, curiosity, contempt, in a word, all the passions of the author are repeated in the lines of his countenance. This imitative picture is only interrupted whilst his proper sentiments, crossing exterior objects, seek for modes of expressing themselves.[5]

Observe particularly the young man caught mid-embodied transparency: strikingly unselfconscious and completely absorbed by the performance. Our direct access to his feelings won't last, of course; to make it more convincing, Siddons stresses its transience. The "youth" is in thrall now, but this spell will be broken any second as his attention wanders off to other "exterior objects."

So when Lovelace manipulates the sentimental assumption that sensitive spectators forget themselves during an affecting performance and allow their bodies show their feelings, his very cynicism proves that this assumption was very much in place in England by the late 1740s. It was in place in spite of the fact that eighteenth-century audience members talked during the play and watched each other as much as they watched the players. Theater, in other words, can serve as a recognizable context for embodied transparency in fiction even in cultures whose traditions of spectatorship differ from ours—which means that it can also serve as a context for faked transparency.

Faking It at the Races

Are there other social settings, besides theater, in which unsuspecting spectators can be turned into objects of observation? It turns out that horse races work very similarly. Writers use races to create situations in which one member of the audience turns away from horses and riders and is struck by the spontaneous body language of another.

Consider the famous scene from Tolstoy's *Anna Karenina* (1877) in which Alexei Alexandrovich Karenin is made to realize how deeply his wife, Anna, is in love with another man, Vronsky, and, worse yet, how incapable she is of concealing her feelings:

> Alexei Alexandrovich was not interested in the race and therefore did not watch the riders, but began absentmindedly surveying the spectators with his weary eyes. His gaze rested on Anna.
>
> Her face was pale and stern. She obviously saw nothing and no one except one man. Her hand convulsively clutched her fan, and she held her breath. He looked at her and hastily turned away, scrutinizing other faces.
>
> "Yes, that lady and the others are also very upset," Alexei Alexandrovich said to himself. He wanted not to look at her, but his glance was involuntarily drawn to her. He peered into that face, trying not to read what was so clearly written on it, and against his will, read in it with horror what he didn't want to know.
>
> The first fall—Kuzovlev at the stream—upset everyone, but Alexei Alexandrovich saw clearly in Anna's pale, triumphant face that the one she was watching had not fallen. When, after Makhotin and Vronsky cleared the big barrier, the very next officer fell on his head and knocked himself out, and a rustle of horror passed through all the public, Alexei Alexandrovich saw that Anna didn't even notice it and hardly understood what the people around her were talking about. But he peered at her more and more often and with greater persistence. Anna, all absorbed in watching the racing Vronsky, could feel the gaze of her husband's cold eyes fixed at her from the side.
>
> She turned around for an instant, looked at him questioningly, and with a slight frown turned away again.
>
> "Ah, I don't care," she all but said to him and never once glanced at him after that.[6]

On the way back from the races Alexei Alexandrovich points out to his wife the impropriety of her behavior, trying to pull her back into his stifling world in which respectable appearances must trump feelings. Anna, who has just witnessed Vronsky's fall from the horse and knows that he is alive but not much more (and who is, moreover, pregnant with Vronsky's child), is unable to do what her husband wants her to: deny everything and laugh at his suspicions with merry indifference. Instead, she announces that she loves Vronsky, that she is his mistress, and that she "fears and hates" her husband. By precipitating this confession, the moment of embodied transparency thus serves as a turning point in the novel.

Still, compared to theater, a hippodrome can elicit only a limited range of emotions from the enthralled spectator. The premise of the situation is that the observer must learn something important about the feelings of the spectator who is watching the horses. And what can the observer learn? He can learn either of two things: that the spectator is in some sort of financial trouble and has a lot riding on the race or that the spectator is deeply emotionally involved with one of the riders. And, anyway, in the case of this second possibility, Tolstoy can be said to have cornered the market. What can be more emotionally engaging than the situation in which the spectator loves the rider and the observer is married to the spectator?

So an interesting development took place in the cinema when film directors adapted many established prose fiction tricks for building contexts for embodied transparency. In films observers still regularly ignore the horses to spy on their fellow spectators and learn something about their feelings, but the observed spectators don't care about horses either. Instead, they often pretend to look at the horses, when, in fact, they are conducting their own business under the assumption that everybody else thinks that they are interested in the horses.

On the one hand this opens up tremendously the range of emotions that the observed spectators can display, because it's no longer about the financial outcome of the races or the well-being of a particular rider. On the other hand people who observe them are now more likely to be quite wrong in their interpretations of these emotional displays, precisely because it's no longer about the outcome of the races or the well-being of a particular rider. One way to describe this situation is to say that this embodied transparency has run away from its context. Horses no longer matter. (Or they matter only to the extent that the spectators think that they matter to others.) People go to races to conduct their affairs and to surreptitiously observe each other.

(Don't misunderstand me. I am not saying that this applies to *all* movies featuring horse races. I am talking only about the movies in which directors use racing to create contexts for embodied transparency: that is, in which there is an observer spying on an unsuspecting spectator. Of course, there are plenty of movies about races in which the focus is on horses and riders and not on spying audience members.)

Think of the famous scene in Ernst Lubitsch's silent film *Lady Windermere's Fan* (1925), in which a group of aristocrats attend the races but nobody pays attention to the horses. Instead, they are preoccupied with spying on other spectators with every possible type of field glass, opera glass, and lorgnette (figs. 6 and 7).

The most important bit of spying, however, takes an unaided eye: Lord Darlington (Ronald Colman) observes at close range the body lan-

FIGURE 6. Spying on other spectators in Ernst Lubitsch's *Lady Windermere's Fan* (Warner Bros., 1925).

FIGURE 7. Mrs. Erlynne (Irene Rich) is framed by binoculars in Ernst Lubitsch's *Lady Windermere's Fan* (Warner Bros., 1925).

guage of his friend Lord Windermere (Bert Lytell) as he appears struck by the sight of the woman who has recently appeared in London and whom nobody knows, one Mrs. Erlynne (fig. 8). Darlington notices Windermere's interest in Mrs. Erlynne, his agitation as other people start gossiping about her, and his attempts to conceal his feelings and appear indifferent. Darlington seems to have caught Windermere at his most transparent, and he infers from what he sees that Windermere and Mrs. Erlynne are lovers. This discovery is important to Darlington because he is in love with Windermere's wife and feels that Windermere's infidelity gives him license to intensify the pursuit of Lady Windermere.

Of course, we know that Darlington is wrong in his interpretation of Windermere's motivation. Windermere is a loving and faithful husband. The reason that he responds to the sight of Mrs. Erlynne so strongly is that at this point he is the only person in their circle who knows that Mrs. Erlynne is his mother-in-law, whose reputation is forever tarnished by her past sexual behavior. To keep that information secret from his wife (who thinks that her mother is virtuous and dead) and the rest of their gossip-mongering friends, he is paying Mrs. Erlynne off. Seeing her now so unexpectedly, and at such close range, and hearing other people speculating

FIGURE 8. Lord Darlington (Ronald Colman) notices Lord Windermere's (Bert Lytell's) interest in Mrs. Erlynne in Ernst Lubitsch's *Lady Windermere's Fan* (Warner Bros., 1925).

about who she might be makes him extremely uncomfortable. In other words even though Darlington is very perceptive about Windermere's body language, we are the only truly appreciative audience for this moment of embodied transparency.

To conceal his discomfort when others start gossiping about Mrs. Erlynne, Windermere pretends to care about the race, busying himself with the program. Similarly, to conceal his "aha" moment when he realizes that Windermere and Mrs. Erlynne have some kind of special relationship, Darlington, too, pretends to think about the horses. He assumes a properly concerned look when Windermere turns to him with his program in hand, matching the other man's pretense with a pretense of his own, peering sympathetically into the program, all the while stealing sly glances at Windermere and enjoying his unconvincing show of interest in horses (fig. 9).

FIGURE 9. Lord Darlington and Lord Windermere feign interest in the program in Ernst Lubitsch's *Lady Windermere's Fan* (Warner Bros., 1925).

Think, too, about the opening of this scene: a large group of men is diligently staring at the horses while Mrs. Erlynne (Irene Rich) is standing in their midst, but when she begins to walk away, they immediately turn and look after her. This sequence shows that they were thinking only of this unaccompanied female stranger in their midst all along. Here and elsewhere in this scene, paying attention to the race is only a pretext for doing something else (fig. 10).

Horse races work the same way in *Notorious*, Alfred Hitchcock's 1946 movie with Ingrid Bergman, Cary Grant, and Claude Rains. Alicia Huberman (Bergman) and Devlin (Grant) are American agents working undercover in Rio de Janeiro to trap a group of Nazis, which includes Alexander Sebastian (Rains). Alicia and Devlin meet at the hippodrome to discuss the progress of their operation. When their conversation becomes too emotional for Alicia (for they also are in love with each other, but

FIGURE 10. Who really cares about races? Men turn to look at Mrs. Erlynne in Ernst Lubitsch's *Lady Windermere's Fan* (Warner Bros., 1925).

Devlin is too ambivalent about his feelings to admit it), she energetically turns her attention to the horses. This is the famous shot in which we can't see her face and only get a view of the race reflected in her binoculars (fig. 11).

And then it turns out that Alicia and Devlin were not the only ones to ignore the horses. All this time, Alex, who is in love with Alicia, was watching her from afar through his binoculars. But, as with other cinematic instances of embodied transparency at the races, his interpretation of Alicia's body language is both astute and wrong. He registers her strong emotions, but he is not sure what they mean. This is why in the conversation that follows Alicia can almost convince him that she "detests" Devlin (a language that would have alarmed a man less blinded by love than Alex):

ALICIA. It was a wonderful race. Did you have much money on the winner?

ALEX. I didn't see the race.

ALICIA. Didn't you? I thought I saw you looking through your field glasses.

ALEX. I was watching you and your friend, Mr. Devlin. I presume that's why you left my mother and me. You had an appointment to meet him.

ALICIA. Don't be absurd. I met him purely by accident.

ALEX. You didn't seem very anxious to get away from him.

ALICIA. Oh, he's just . . .

ALEX. I watched you. I thought maybe you're in love with him.

ALICIA. Don't talk like that. I detest him.

ALEX. Really? He's very good-looking.

ALICIA. Alex, I've told you before. Mr. Devlin doesn't mean a thing to me.

ALEX. I'd like to be convinced. Would you maybe care to convince me, Alicia, that Mr. Devlin means nothing to you?

FIGURE 11. Horse races in Alfred Hitchcock's *Notorious* (RKO, 1946).

To "convince" Alex, Alicia has to marry him. Their marriage sets into motion the successful spying campaign against Alex's Nazi friends, antagonizes Alicia and Devlin, and leads to Alicia's near death at the end of the movie and Alex's certain death once the credits come onscreen. A horse race is thus a setting that often turns around the lives of spying movie protagonists. It's where they *think* they learn some important truths about the people whom they observe, while, in fact, it's where they come up with wrong interpretations that serve to thicken the plot.

There is a lovely ironic disconnect, in other words, between the world that the unsuspecting characters *think* they live in and the world the movie directors spring at them. In the first world horse races are still what they used to be in *Anna Karenina*: a straightforward setting for moments of embodied transparency. Hopeful Lord Darlingtons and Alexander Sebastians put on their hats and set off to the hippodrome to catch others at their unguarded moments. But when they arrive there, the hippodrome turns out to be a place where everybody is busy figuring out the angles, scheming, and pretending. Only no one warned them about it. So they proceed to diligently observe other people's body language. Instead of reading minds, however, they end up misreading them (which is just what a good story needs).

Soccer: The Last Bulwark

But soccer is different. Cinema may have corrupted horse races but not athletic events. The stadium remains a reliable source of embodied transparency. Characters still forget themselves when they watch football (i.e., British soccer), basketball, hockey, and other competitive sports, so a strategically placed observer can get an undiluted shot of direct mind reading.

Take Nick Hornby's *Fever Pitch* (1992)—an autobiographical memoir of a football fan. The very first game that the protagonist attends, at the tender age of eleven, the game that, in Hornby's scheme of things, rechannels his unhappiness about his parents' divorce into his obsession with the team Arsenal—a fateful game!—is described in *Fever Pitch* as an occasion for direct access to other people's minds. As the narrator puts it,

I remember looking at the crowds more than at the players. . . .

. . . What impressed me most was just how much most of the men around me *hated, really hated* being there. . . .

I'd been to public entertainments before, of course; I'd been to the cinema and the pantomime and to see my mother sing in the chorus of the *White Horse Inn* at the Town Hall. But that was different. The audiences I had hitherto been a part of had paid to have a good time and, though occasionally one might spot a fidgety child or a yawning adult, I hadn't ever noticed faces contorted by rage or despair or frustration. Entertainment as pain was an idea entirely new to me, and it seemed to be something I'd been waiting for.[7]

I didn't emphasize *"hated, really hated"* above—Hornby did—but I would have if he hadn't. What this phrase shows is that the eleven-year-old protagonist can see what men in the crowd *really* feel: rage, despair, and frustration. One doesn't come across such rich fare at the theater, where an honest yawn is a rare treat. Attention, piety, appreciative laughter, thoughtful sadness, or boredom can all be put on to impress other spectators, but try faking unremitting pain and hatred!

Though if you want to see bodies truly possessed by uncontrollable emotions, wait until Arsenal wins. At one point the narrator, now fully grown and fully addicted, is so overwhelmed by the unexpected triumph of his team that he is acting out his joy not knowing what he is doing, while the fans of the losing team are watching him: "It was the second of three or four lifetime football moments where my delirium was such that I had no idea what I was doing, where everything went blank for a few moments. I know that an old man behind me grabbed me around the neck and wouldn't let go, and that when I returned to a state approaching normal consciousness the rest of the stadium was empty save for a few Tottenham fans who stood watching us, too stunned and sick to move" (181).

What's striking about this scene is how embodied transparency hits everyone: the observers and the observed. If the narrator does not know what his body is doing—if he is in such a state that he doesn't have any mental picture of himself—this means that the Tottenham fans not only have direct access to his mind via his body language but that they also will always know more about him at that moment than he, or his readers, can know. That is, the observers know what Hornby did when his mind went blank, and nobody else does. But at the same time, the narrator, once he comes to, can observe the body language of the Tottenham fans, who are quite transparent to him in their numbing despair. He knows they watch him not because they enjoy the spectacle of Arsenal fans' victorious delirium but because they are "too stunned and sick to move." One wonders to what degree they are aware of *their* body language just now, for the narrator's temporary lack of self-consciousness is mirrored by their own.

And then there is also the neck-grabbing old man, whose body language makes him transparent to Hornby even as Hornby loses the awareness of his own, and who must be unselfconscious about his actions while being wrapped up (so to speak) in Hornby. Embodied transparency sweeps over the fans like a storm: even as they see what it does to others, they can't control what it does to them.

I find it significant that both the book and the feature film based on it (the British version)[8] turn to these moments of direct mind-access to explain the magnetism of the game, even though their explanations differ. Hornby thinks that one reason he needed soccer so badly was that the palpable pain of Arsenal supporters jived with his own depression. Or, as he puts it, "I go to football for loads of reasons, but I don't go for entertainment, and when I look around me on a Saturday and see those panicky, glum faces, I see that others feel the same" (135–36).

I agree that in darker moods one sometimes appreciates being around other unhappy people. But note a particular type of mind reading taking place here. The therapeutic times of shared anguish also happen to be the times when Hornby knows—or believes he knows, which is the same—what other people around him are feeling. These are moments of intense mind-reading pleasure. (Real stuff, too: not just actors onstage going through rehearsed paroxysms of passion.)

To marvel at the irresistible pull of the game, the movie *Fever Pitch* takes up as its point of departure a passage from the memoir in which Hornby talks about the spectators' "absorption" in what's happening in the field.[9] Observe as you read that passage, below, how the verb tense changes from past to present once we are inside the moment of embodied transparency, intensifying the impression of immediacy and transience:

We beat Everton 3-1 that night, 4-1 on aggregate, a comfortable enough win which Arsenal fully deserved, but we had to wait for it. Four minutes before half-time Rocastle beat Everton's offside trap, went round Southall, and stroked the ball well wide of a completely empty goal; and then three minutes later Hayes was through too, only this time Southall brought him down six inches from the goal line. Hayes took the penalty himself, and, like McClair, booted it well over the bar. And the crowd is going spare with frustration and worry; you look around and you see faces working, completely absorbed. (198)

In the movie the protagonist, Paul Ashworth (based on Hornby), played by Colin Firth, sits home alone, watching a replay of an old match (fig. 12). He has just had a falling-out with his girlfriend, who is losing patience over his obsession with the game, and he is trying to understand what it is about football that draws him in so powerfully.

We hear his thoughts in a voice-over when he as good as tells us that the most magical part of the game is the feeling of direct access to other minds that it seems to induce: "What about this? Three minutes to

FIGURE 12. Colin Firth as Paul Ashworth in David Evans's *Fever Pitch* (Channel Four Films, 1997).

go and you are 2-1 up in a semifinal. You look around and see all those thousands of faces contorted with fear, and hope, and worry. Everyone lost. Everything else gone out their heads. Then the whistle blows, and everyone goes spare. For those few minutes you're at the center of the world."

You are at the center of the world for those few minutes when you feel that you know what everybody else around you is thinking. It's mind reading on a massive and yet intensely personal scale because your own feelings are reflected in the feelings of others. Their bodies are transparent to you, but so is yours to them, and yours to you through its reflection in theirs. The experience is brief, too, though these "few minutes" may feel longer or shorter, depending on what's happening on the field.

It seems, then, that at this point our cultural imagination still treats sporting events as a reliable venue for direct access to spectators' minds. Theater stopped being such a venue a long time ago, with the advent of the novel, and so did, more recently—with the advent of cinema—horse races. But both real-life and fictional football fans can still look at other people around them in the stands and see strong feelings written large on their bodies while "everything else [has] gone out their heads." No performance here, no trying to impress others one way or another.

Most likely, this state of affairs won't last. The culture of greedy mind readers subverts contexts for reliable transparency as soon as it becomes fully aware of them. Perhaps it's only a question of time before we see a movie in which a modern-day Lovelace brings his Clarissa to a football game and fakes his emotions, hoping that she will attend to them instead of the events on the field. And, when that happens, who knows what will emerge as the next social setting in which being a spectator means being a show: passionate and transparent at the same time?

FIVE

In which readers return to Hitchcock's *Notorious* and realize that neither a bowl of soup nor a cup of coffee are what they seem to be / Matt Damon plays poker / Humphrey Bogart loses the girl / Helen Mirren loses a crumpet / Ralph Fiennes loses his cool / Hillary Clinton almost coughs / and blame is wrongly assigned.

Movies

The Power of Restraint

We are now moving away from the printed page and toward visual and aural experiences of embodied transparency: mostly in movies, but with occasional glances at the theatrical stage.

Movies, of course, *are* theory-of-mind-writ-large—competing with novels as feasts for greedy mind readers. To date I am aware of two excellent studies of theory of mind and feature films—Per Persson's *Understanding Cinema: A Psychological Theory of Moving Imagery* (2003); and Colin McGinn's *The Power of Movies: How Screen and Mind Interact* (2007)—though this topic is so rich that I expect more books dealing with it to appear soon.[1]

What I do in this chapter by no means adds up to a comprehensive analysis of cinema and theory of mind; my focus is intentionally narrow. McGinn writes that onscreen "the eyes become liquid pools of dense feeling. It is as if we are *seeing* the emotions of the characters, so entwined are the images and the feelings (at least when the movie is doing its job)."[2] I start therefore with the assumption that if the movie is doing its job, it is already awash in embodied transparency. Still, not all transparency is created equal. Even in this saturated medium some moments of access feel richer than others, an effect achieved by selectively obscuring the emotions of the characters. In what follows I look at one particular pattern of such strategic obstruction: moments when characters attempt to conceal or restrain their feelings and, by doing so, become more interestingly transparent than their freely emoting counterparts.

Beyond the Kuleshov Effect

What makes movies particularly conducive to embodied transparency is a combination of two factors: theory of mind and montage. We see a face on the screen and we are ready to assume (because our theory of mind never stops working!) that there is something on the person's mind, and the technique of montage tells us what that something is. You may remember the famous experiment from the early days of cinema in which the Russian filmmaker Lev Kuleshov alternated the shot of an actor wearing the same neutral expression with the images of a bowl of soup, an old woman in a coffin, and a little girl. In each case audiences reported seeing a different emotion on the man's face: hunger following the shot of soup, sadness following the shot of the coffin, and happiness following the shot of the little girl.

The "Kuleshov effect" seems to imply that movie actors don't have to do anything at all, while all that directors have to do is provide sequences of images juxtaposed with faces; audiences will supply the emotions and thus the meaning of each scene. But while there is no question that cinema exists because of our readiness to read emotions into body language, the stories that movies actually tell are much more complex than a "story" born out of the juxtaposition of a man's face and a bowl of soup. Consider the shot of Alicia's face followed by a shot of a cup of coffee from Hitchcock's *Notorious* (figs. 13 and 14). If we take it to mean that she is thirsty, we're wrong. She is struck by the realization that her husband and mother-in-law are slowly poisoning her and that the cup of coffee in front of her contains another dose of that poison. To be able to interpret Alicia's state of mind correctly and to grasp the meaning of the scene, we need the whole background narrative with its complex web of mind attributions: Alex and his mother have known for some time that Alicia is an American agent, but they can't let other people in their Nazi circle find this out because if they do, they'll kill Alex; Alicia has just realized that they know who she is and are determined to do her in, gradually and quietly. They are desperate, and she is in their power. In fact, she is *completely* in their power because if Devlin does not love her—and she begins to believe that he doesn't—he won't come to her rescue. As we sense Alicia's terror, *Notorious* begins to feel like a horror movie.

Directors thus both use the Kuleshov effect and go beyond it. They count on our reading an emotion into a face juxtaposed with a shot of an object, but they also make sure that to read that face correctly, we need much more than just this immediate sequence of shots: we need information about everybody's thoughts and feelings prior to this particular scene. Thus at any given point in the movie what drives our interpretation of a character's mental state are our earlier interpretations of other mental states.

Of course, keeping all these interpretations in mind and applying them to the present sequence of shots implies quite a bit of cognitive effort and uncertainty. We are way beyond the simple equation: a bowl of soup + a man's face = the man is hungry. We know what one character thinks, we are not quite certain what another thinks, and we have no clue what the third thinks. For instance, does Devlin love Alicia? We suspect that he does, but we won't find out for sure until the very end of the movie. Or, do Alex and his mother realize, as Alicia first looks at the cup of coffee and then glances up at them, that she is now aware of their plan to gradually poison her? We suspect they do, but it remains ambiguous.

By introducing doubt and ambiguity into our interpretation of characters' mental states, directors create onscreen versions of real-life social complexity. This means that when they grant us our "aha!" moments— that is, when they make us feel that we know exactly what the characters think—we appreciate it much more than we would have if the characters had been transparent all the time. The moments of occasional complete access make us feel like brilliant social players. We cherish these illusions of superior social discernment and power because they stand out both amidst our daily mess of mind-reading uncertainty *and* amidst the uncertainty carefully constructed by film directors. (And of course certain cinematic genres make mind-reading uncertainty their raison d'être; in detective movies everything is geared toward preventing us from reading characters' minds for as long as possible.)[3]

FIGURE 13. Alicia Huberman (Ingrid Bergman) looks at a cup of coffee in Alfred Hitchcock's *Notorious* (RKO, 1946).

The shot of Alicia's face as she looks first at the cup of coffee and then at her husband and mother-in-law is an instance of embodied transparency. So is, a moment earlier, Alex's worried exclamation as he stops his Nazi friend, Dr. Anderson, from accidentally picking up and drinking Alicia's poisoned coffee. Once you start scrutinizing movies for such moments of perfect access, you realize that they occur often enough to make watching a movie a rather extraordinary mind-reading experience. You also notice that restraint—which, as you may remember, is one of the three "rules" for embodied transparency in the novel, though not, perhaps, the most important one—assumes new importance on the screen. It seems that directors intuitively but persistently look for social contexts that allow characters to struggle to rein in their emotions.

Why should it be so? What's so special about restraint on the screen?

FIGURE 14. *Notorious*: the coffee cup.

Royal Path to Embodied Transparency

Humphrey Bogart told this story: When they were shooting *Casablanca* and S. Z. (Cuddles) Sakall or someone comes to him and says, "they want to play the 'Marseillaise,' what should we do?—the Nazis are here and we shouldn't be playing the 'Marseillaise,'" Humphrey Bogart just nods to the band, we cut to the band, and they start playing "bah-bah-bah-*bah*."

Someone asked what did he do to make that beautiful scene work. He says, "they called me in one day, Michael Curtiz, the director, said, 'stand on the balcony over there, and when I say "action" take a beat and nod,'" which he did. That's great acting. Why? What more could he possibly have done? He was required to nod, he nodded. There you have it. The audience is terribly moved by his simple *restraint* in an emotional situation—and this is the essence of good theater.

—Mamet, *On Directing Film*

Mamet turns to the Bogart anecdote to illustrate his larger argument about what separates good acting from overacting. Instead of trying to portray emotions, actors should perform what he calls "uninflected" physical actions. A simple nod goes a long way because—remember the Kuleshov effect?—viewers read emotions into it based on the context of the scene.

I wonder, though, how much we should make of Mamet's saying that restraint in an emotional situation is "the essence of good theater" while describing a scene from a movie. Most likely he means "theater" broadly: as any performance, whether it takes place onstage or onscreen. But perhaps he does mean the theatrical stage, and his choice of example unintentionally reflects the fact that cinema may have more use for restraint than theater does.

Restraint is an emotion that can be detected and appreciated if we look at the body closely. Both theater and cinema cultivate this kind of attention, but cinema can resort to crude force not available to theater. A close-up leaves us no choice but to attend to the character's face, while in theater there is always a chance that spectators are looking somewhere else while a character pointedly does *not* move a muscle when one would expect him at least to blink. (Not to mention that if we are seated far away from the stage, we simply can't see the facial expressions of actors, and the fine points of restraint are lost on us.)

Note, too, that when Mamet talks about that scene in *Casablanca,* he doesn't worry about distinguishing between the actor and his character. When he says that the "audience is terribly moved by his simple restraint in an emotional situation," the pronoun *his* refers to Humphrey Bogart as much as to his character, Rick Blaine. This is quite understandable. As a director Mamet is concerned about the behavior of actors and the effect of their behavior on the audience. Bogart shows restraint by not overacting, indeed, by almost not acting at all—"he was required to nod, he nodded"—and the audience responds well to his insightful portrayal of the emotional situation.

For the purpose of our discussion, however, I must distinguish between the actor and the role and ask why the audience should be moved by the restraint exhibited by a character. After all, it's not like there is some absolute value placed on self-control in our culture, so that every exhibition of restraint is recognized and applauded as a sign of virtue. A stiff upper lip is appreciated by some people in some situations and disapproved of by others. Moreover, there are many other moments in *Casa-*

blanca when, instead of showing restraint, Rick displays strong emotions (as, for example, when he strikes the table with his fist and says about Ilsa's sudden reappearance, "Of all the gin joints in all the towns in all the world she walks into mine"). Such scenes are moving and memorable, and we hardly hold it against Rick when he lets it all out.

I suggest, then, that we like it when characters show restraint, not because restraint is good in and of itself but because restraint may be used as a means for *interestingly complex* embodied transparency, more complex, for instance, than the one exhibited by Alex in *Notorious* when he gasps and stops Dr. Anderson from drinking the poisoned coffee, or even by Rick when he strikes the table with his fist. What is fascinating about restraint in the movies is that often we don't even have to know what particular feeling the character is trying to restrain—*what is transparent is the struggle for self-control.*

For instance, we don't know exactly what combination of pride, defiance, hope, personal grief, and recklessness Rick is experiencing when he nods to the orchestra to let them play the "Marseillaise," but we do know for sure that he is controlling himself by not saying anything or expressing any of these feelings directly. So he is transparent to us in his complexity, flattering, as it were, our theory of mind with a promise of access to not just one emotion but to a whole suite of emotions.

Restraint, thus, is a royal path to embodied transparency in the movies because it can be superimposed on any emotional content, including that which is not completely accessible to us. I said earlier, in my discussion of embodied transparency in the novel, that restraint allows for displays of appealingly complex mental states, as in, "I don't want her to know what I am feeling." Restraint on the screen allows for the same third-level cognitive embedment, yet on top of it, it can also visibly expand the palette of feelings that the character is endeavoring to conceal. This third-level embedment becomes a third-level embedment with a flourish, as it were: "I don't want her to know what I am feeling, but I am feeling an awful lot." Hence both the novel *Pride and Prejudice* and its movie version tell us that Mr. Darcy tries to restrain his anger when Elizabeth turns down his marriage proposal, but we can read more than that into Colin Firth's expression (see fig. 3). Once the actor's face gets factored into the equation, our mind-reading adaptations have something extra to process.

It's not accidental that the "neutral" face in Kuleshov's experiment was that of the actor Ivan Mozzhukhin, known as the "Russian Valentino." Since its early days the cinema has sought out actors who are particularly good at conveying the impression of complexity behind restraint. As Anthony Lane writes in his review of the movie *Biutiful,* "Bring me the head of Javier Bardem. Did you ever see a nobler noggin? . . . The scene in which [Bardem's character] finds [his son] alone, abandoned by his mother, with a bruised face, is deeply upsetting, and all the harder to erase because Bardem plays it so calmly, reigning in the urge to erupt. Dormant volcanoes, ready to rumble, are always the ones to watch."[4] The put-on calmness offers our theory of mind a delectable combination of complexity and access. (And certain movies are only too happy to turn our theory of mind into a gourmet.)

Photogenic Professions

Nowadays, people want glamour and tears, the grand performance. I'm not very good at that. I've never been. I prefer to keep my feelings to myself. And foolishly I believed that that's what people wanted from their Queen. Not to make a fuss, nor wear one's heart on one's sleeve.

—Queen Elizabeth II, *The Queen*

A person whose profession calls for the exercise of restraint makes for an interesting movie character. Doctor, lawyer, spy—anybody whose job involves dissociating herself from the immediate emotional content of a situation and thus suppressing her feelings—is excellent material for embodied transparency. And it is even better if she happens to be personally invested in the situation at hand. Oh, the passions behind that calm facade! Cinema thrives on this kind of inner conflict.

Stephen Frears's movie *The Queen* (2006) goes for the ultimate case of professional restraint. Frears's protagonist, Queen Elizabeth II (played by Helen Mirren), suppresses her emotions around the clock. The decorum associated with royalty precludes her from wearing her heart on her sleeve, and she comes to love it. Restraint becomes her: Elizabeth takes pride in the "quiet dignity" with which she comports herself. She believes that this is what "the world has always admired her for." When, after

Princess Diana's death, the British people demand that their queen grieve and display her emotions for public consumption, she is shaken.

She is shaken—but will she show it? The movie goads us with the promise of the queen's emotional display. Other people around her—Tony Blair, Cherie Blair, Prince Charles, and Prince Phillip—let us see their anger, impatience, fear, surprise. Will she?!

I said before that restraint is a royal road to embodied transparency in cinema. Now make that a royal restraint—literally—and, as a director, you have a recipe for a very special treat for our mind-reading adaptations. When the position of a protagonist is such that she must control and subdue her emotions not just from eight to five but constantly, you can build an entire movie around different shades and modes of restraint. This is another way of saying that you can pick and choose you moments of embodied transparency and pile them up as thickly as you want.

Frears does precisely this. For instance, several times in the course of the movie, just as she is about to start eating, the queen is called to the phone for an urgent and stressful conversation with Tony Blair about the Diana crisis. At least on one such occasion, the queen has already opened up her mouth to bite off something distinctly yummy, and there is that phone call again. The queen restrains her emotions, of course, but we can imagine how she feels as she gives a brief parting look to the chocolate cake before leaving the room (fig. 15).

A different show of restraint occurs when the queen learns from her adviser that a ceremonial procession known under the code name Tay Bridge will be used for Diana's funeral. Tay Bridge was conceived and rehearsed for the future funeral of the Queen Mother, so Elizabeth must be struck with the impropriety of lavishing it on a disgraced daughter-in-law, who is not even considered a member of the royal family anymore. But, however shocked she is, she suppresses it. All she does as her secretary tells her about Tay Bridge is say quietly, "Anything else?" and briefly touch her forehead.

The queen must exercise restraint even with her nearest and dearest. At one point she is driving the car, with her son Charles in the passenger seat, getting ready to stalk a deer. (At the royal estate in Balmoral they have to drive to the place of the hunt.) Charles keeps talking about how warm Diana was, how much the British public loved her and hated the rest of the royal family, and how afraid he is now of assassination. The

FIGURE 15. The queen (Helen Mirren) gives up her crumpet in Stephen Frears's *The Queen* (Pathé, 2006).

queen must be getting terribly annoyed, but she controls herself. All she does is observe cheerfully that she has just changed her mind, that she does not want to go deer stalking with Charles and the rest of the family, and that she will walk with her dogs instead. Then she gets out of the car.

Even when she is alone, Elizabeth cannot relax for long. When during a solitary ramble in Balmoral, her car breaks down as she drives through a shallow creek, and she is waiting for help to arrive, she begins to cry, worn out by the stress following Diana's death and the pressure put on her by the media and her advisers (fig. 16). But even there, with nobody to observe her and appreciate her fabled "quiet dignity," she cries with her face turned away from the camera (that is, from us).

Just as she begins to turn toward us, a stag walks onto the shore, and Elizabeth is struck by its beauty. She looks at it first with deep admiration and then with intense alarm as she hears guns firing and dogs yelping, the deer-stalking party approaching. She shoos the stag away and then just stands there for several moments, taking in the river, the mountains, the sky, suddenly happy again, rejuvenated by the encounter. But then, once more, she feels compelled to restrain her emotions. She wipes her eyes and nose with a scarf and composes her features into an impenetrable, severe mask, ready to encounter the world as the imperturbable, dignified Queen Elizabeth.

It's a spectacular scene, perhaps the only time in the movie when the mask is completely off and we see the queen going through a succession of emotions. By now Frears has already trained us to be grateful for crumbs—for expressions of feeling so subtle as to be invisible—so we perceive this display as lavish and exceptional.

In other words restraint exercised by the queen throughout the story works in two ways. First, on the occasions when she lets go of it and shows her emotions freely (as in the deer scene), it feels rare and valuable. Second, restraint creates an illusion of privileged access to complex emotions. For instance, the brief wistful look that the queen casts at the chocolate cake just before being whisked away to take Blair's phone call momentarily conjures up an image of a disappointed child deprived of a treat. Young children can be embodied transparency personified, but this particular moment of *restrained* disappointment promises an opening into a whole gamut of complex feelings. Of course we don't see these feelings, but our theory of mind, all fired up by the present show of restraint and the previous information about the queen's ambivalent attitude toward Blair and Diana, keeps us guessing what they might be.

FIGURE 16. The queen cries with her back to the camera in Stephen Frears's *The Queen* (Pathé, 2006).

Viewed from the cognitive perspective, the movie is but a sequence of embodied transparencies, each making our theory of mind buzz in a slightly different way. The transparency exhibited by freely emoting characters (such as Tony Blair, Cherie Blair, Prince Phillip, and, presumably, formerly, Diana) feels different from the one snatched from the queen when she thinks nobody's watching and different again from the one snatched from the queen when she thinks someone is watching. It only makes sense that the medium that supplies greedy mind readers with nonstop fantasies of access has to develop different strategies for delivering its "aha" moments.

Poker Faces

Doctors, lawyers, spies, and British queens are not the only ones who make promising movie characters because their line of work calls for the exercise of restraint. So do gamblers.[5] Cinema loves poker faces— that is, not real-life faces that show no emotions whatsoever; who wants those?!—but faces of people playing poker. We have long poker sequences—and thus displays of poker faces—in movies including *Ocean's Eleven, The Sting, Casino Royale, Cassandra's Dream,* and *Rounders* (fig. 17).

The game of poker is particularly good for a movie because it comes with several layers of restraint. First, players must remain dispassionate during the game in order to conceal their true positions from others. Second, frequently in the story more is riding on the outcome of the game than just the pile of money: the protagonist's whole life will take a certain turn depending on whether he loses or wins. (If he wins, for example, he will be able to stay with the girl of his dreams, he won't have to murder somebody he was contracted to murder for a hefty fee, etc.) So as he sits there at the card table, looking blandly indifferent, we know not only that he is waiting with bated breath to see who will get the cash but also that he feels agonizingly suspended between his two lives, passionately wishing for one, mortified at the thought of the other.

And, again, because restraint can be superimposed on a variety of emotional contents (anger, grief, joy, painful uncertainty, disappointment)—we don't need to know immediately whether the protagonist is

FIGURE 17. Mike McDermott (Matt Damon) at a card table in John Dahl's *Rounders* (Miramax, 1998).

bluffing. We do know that whatever his feelings are, he labors to conceal them, and that's the extent of his transparency until the end of the game. Of course, when the game is over and we do find out if he was bluffing, this retroactively deepens our impression of access. We can then look back at his behavior at the table and develop a more nuanced perception of fear and hope that must have consumed him as he sat there with a stony expression (or affected to appear serene and relaxed).

That's what movies (and poker reality shows) do. In May 2011, however, the American media got an opportunity to construct a plausible narrative of restraint and embodied transparency (not in these terms, of course) based on real-life historical events. Shortly after the killing of Osama bin Laden, journalists began commenting on the "poker face" that President Obama presented to the world as the secret raid on bin Laden's compound in Pakistan was about to begin and on the "stone face" that he maintained later, while receiving updates as the actual raid unfolded in Abbottabad.

Thus writing about the White House Correspondents' dinner, at which Obama joked about Donald Trump and other people who wanted to see his American birth certificate, the *New Yorker*'s David Remnick insisted that the "truly astonishing aspect of the dinner was not the political japery but Obama's knowledge that, as soon as the weather in northern Pakistan cleared, his own black helicopters would ferry a crew of Navy SEALs to bin Laden's compound in Abbottabad."[6] Maureen Dowd of the *New York Times* observed how perfectly the "president's studied cool and unreadable mien" served him on this occasion. *Salon*'s Peter Finocchiaro actually titled his short piece "Obama's Poker Face," posting a video of Obama laughing merrily at Seth Meyers's joke about bin Laden's apparent invulnerability, leaving it to us to arrive at the inevitable conclusion that the president's laughter was covering some very specific thoughts about bin Laden's impending fate.

The *New York Times* article "Behind the Hunt for Bin Laden" provided memorable snapshots of embodied transparency involving both Obama and his secretary of state, Hillary Clinton. It quoted one of Obama's aides as saying that while the president and his national security team were receiving updates on the raid, "Mr. Obama looked 'stone faced.' "[7] Although the accompanying photo showed Obama looking intense and focused and not at all impassive, the aide's comment about the president's stone face—which, given how much was at stake during the thirty-eight-minute raid, *must have* concealed intense emotions—was immediately picked up by news outlets in America and around the world.

And so was the gesture of Hillary Clinton from the same photo. Clinton is shown clasping her hand to her mouth, which can—and has been—interpreted as reacting emotionally to what she is hearing and trying to restrain her emotions. Although later she said that it's likely that she had been merely "preventing one of her early spring allergic coughs,"[8] more *interesting* interpretations, all evoking embodied transparency (e.g., that her gesture betrayed "shock" even as she wanted to restrain herself) proliferated in the press both in America and abroad.[9]

True, responses to Obama's and Clinton's body language feed stereotypes about gender and about the personal styles of these politicians. It's also true that a photo op arranged on what its subjects must have known would be considered a historic occasion (no matter what the outcome of the raid) was unlikely to capture any authentic unpremeditated gestures. But, putting these considerations aside for a moment, the media-fueled obsession with Obama's "stone face," or "poker face" *(which is assumed to have been concealing strong emotions)*, and Clinton's purported barely restrained "shock" shows how irresistibly fascinating transparent body language is, especially if it can be perceived as transpiring in "real life."

Refusing to Watch

Emoting movie characters are not the only ones who can exhibit restraint and, by doing so, make our illusion of access more exciting. The same effect is achieved when other characters refuse to watch an individual who is experiencing strong emotions and the camera moves away from him or her at a crucial moment. The assumption behind this strategy is that sometimes people's faces expose their feelings to such a degree that they become painful to watch, unless a person who watches has a sadistic streak. It is transparency by omission: we can't see the actual face, but our imagination magnifies its emotional nakedness. Robert Redford, director of the movie *Quiz Show* (1994), uses this strategy at a crucial moment both to intensify our impression of one character's transparency and to make more sympathetic another character, who refuses to enjoy that show of transparency.

It's almost inevitable that Charles Van Doren (Ralph Fiennes), the protagonist of *Quiz Show,* plays poker and is good at it. Charles's behavior during the game—he is relaxed, smiling, alert, inscrutable, and bluffing—models perfectly his behavior throughout the movie: he is charming, inscrutable, and lying. The film tells the story of the 1950s scandals surrounding television quiz shows such as *Twenty One* and *Tic Tac Dough.* An attractive instructor from Columbia University and scion of a prominent intellectual family, Charles Van Doren becomes the champion of *Twenty One* after dethroning another longtime winner, Herb Stempel (John Turturro), a "fat, annoying Jewish guy with a sidewall haircut," presumably less palatable to the show's gentile corporate sponsors.

Van Doren's rise is spectacular because the show is rigged. Stempel, Van Doren, and other champions who come before and after them are given answers beforehand and told exactly what to do on the air: how to behave to ratchet up the drama and whether to continue winning or to "take a dive." (This is essentially protoreality TV, staged carefully to maintain the effect of immediacy and spontaneity. Stempel and Van

Doren *could* perform on their own—they are well informed and widely read—but the producers are putting on a show, which means leaving nothing to chance, scripting and intensifying every emotional moment.) Charles dislikes this deception, but once he is in, he upholds it. And his poker face serves him well.

Still, the moment comes when, like the queen in Frears's movie, Charles loses his cool. Forced to appear before a congressional committee investigating quiz shows, he has to tell the truth about having been supplied with answers and coached in his "spontaneous" intellectual breakthroughs and emotional reactions. Painful as his testimony is—with indignant reactions from the committee members and the attending public—Charles manages to stay relatively composed. But when he leaves the room, accompanied by his parents, who are clearly shattered by these revelations but still supportive of their son and struggling to maintain *their* composure, he is attacked by journalists, who are determined to get under this patrician family's skin.

First they ask Charles if he knows that he has just been fired from NBC, where he had been promised a plum spot on an educational show, once his stint with *Twenty One* was over. Charles is taken aback, but still he says calmly that, no, he didn't know that. The journalists then turn to his father:

JOURNALIST. Professor Van Doren, are you proud of your son?
FATHER. I've always been proud of Charlie.
ANOTHER JOURNALIST. Proud of what he did?
FATHER. The most important thing now is for Charlie to get back to his teaching.
ANOTHER JOURNALIST. Did you know that the Columbia Trustees are meeting right now? They're going to ask for Charlie's resignation.

At this point Charles Van Doren's mother closes her eyes to hide her pain. The journalists have apparently succeeded in reducing her to pure anguish. But the father is still standing, so they focus on him again:

JOURNALIST. Professor Van Doren, you spent your whole career at Columbia. What's your reaction to that?

The older Van Doren, speechless, rubs his temple. The journalists persist:

JOURNALIST. Professor Van Doren?

Something important happens then. Herbie Stempel, the previous champion of *Twenty One,* is also there in the hall. The present hearings have vindicated his earlier unsuccessful attempt to convince the congressional committee that the show was rigged. Herbie has every reason to dislike Charles and rejoice in his fall from grace. When the journalists set upon the Van Dorens, Herbie watches Charlie's face eagerly (fig. 18). But just as they pounce again on the older man—"Professor Van Doren?"—Herbie can't take it any longer and steps behind the corner so as not to see the painful scene.

Aware of his father's acute suffering, Charles Van Doren manages to say, "Dad, go ahead with mother. I'll meet you later." "No reaction," observes a journalist (satisfied, we presume, with bringing low yet another Van Doren) and redoubles his attack on the main victim: "Charles, a few more questions." Just then another journalist spots Herbie around the corner. (fig. 19).

JOURNALIST. Herb Stempel. Herbie, how 'bout a picture—you and Van Doren together.
HERB. No, not now. Christ, look at the guy.
JOURNALIST. Come on, the both of you.

FIGURE 18. Herbie (John Turturro) watches Charles's (Ralph Fiennes's) face eagerly in Robert Redford's *Quiz Show* (Hollywood Pictures, 1994).

FIGURE 19. Herbie steps behind the corner in Robert Redford's *Quiz Show* (Hollywood Pictures, 1994).

HERB. You know what the problem with you bums is? You don't leave the guy alone unless you're leaving him alone.

This is our last glimpse of Herb Stempel. The camera returns to Charles. The journalists are closing in on him: "Who do you blame, Charlie?" "Do you feel that the committee has treated you fairly?" "How is the pressure compared to *Twenty One?*" But just as it begins to look like Charles is about to break down, the camera moves back to the room where the producers of *Twenty One* are testifying before the congressional committee.

In other words, we do not actually get to see the emotional disintegration of the protagonist. When Charles loses it—that is, if he loses it: we will never know for sure—the camera is strategically somewhere else: first with Herb Stempel, then with the congressional committee. This is the same tack that Frears uses in *The Queen* when Elizabeth cries with her face turned away from us. Embodied transparency is strongly implied but not shown.

One can argue that by not letting us see the protagonist's face at this crucial moment, the directors follow the ancient theatrical principle of taking the tragedy offstage to make the viewers' imagination work harder. Just as murders and suicides may seem more striking when we don't actually see them, so imagining Charlie when his poker face finally crumbles might be a more powerful emotional experience than actually seeing Ralph Fiennes's expression at this point. When Herbie says to the journalist, "No, not now. Christ, look at the guy," we must shudder at what we would see were we to "look at the guy."

But, apart from preventing us from looking at Charlie, Herb Stempel's reaction accomplishes something else. Until this moment Herbie was coming across as a bit of a schmuck: he lied and made stupid business investments; his social skills seemed barely adequate, his emotional responses crude or naive. But his present decision not to watch Charlie's face complicates all that. It reveals some essential nobility of his character, and since this is our last encounter with Herbie, this perception stays with us after the movie is over.

Embodied transparency can be used to communicate to us something important about the people who witness it. By saying this, I am expanding the argument that I made in chapter 2, talking about Mr. Darcy's trying—and failing—to conceal his anger at Elizabeth Bennet's rejection, and Elizabeth's discomfort during that "dreadful" pause.[10] There I suggested that if you are a writer and you want your character to remain sympathetic, you don't put her in a situation in which she begins to enjoy the spectacle of someone else's transparency, thus coming across as sadistic. Here I am turning to situations in which characters are acutely aware of their position as witnesses of other people's transparency and choose to do something about it.

Herb Stempel's choice is to consciously remove himself from the position of a privileged observer. His actions are particularly appealing because they represent a stark contrast to the behavior of the journalists, who take pictures of Charles Van Doren and his parents while aggressively forcing them into a state of transparency. Of course, we don't have to think of their sadism as intentional and thus indicative of some major character flaw. We can say instead that they are only "doing their job." Still, Herbie's repeated refusal to go along with them—first by instinctively stepping behind the corner and then by categorically rejecting a photo op with Charles, even though that would have brought him renewed publicity—marks him as a man of a finer moral caliber precisely because *he* would not do his job anymore.

For creating a spectacle of embodied transparency used to be his job, too. Participating in *Twenty One* meant pretending to be on the spot: visibly sweating in search of an answer, visibly rejoicing if the answer is correct, visibly despairing if it is wrong. Herb Stempel used to be part of a world in which embodied transparency is a valuable commodity—the world that digested Charles Van Doren, the world in which the journalists feel it is their right and their duty to wring transparency out of their subjects—but he is not of that world anymore. In the moral economy endorsed by this movie, a refusal to buy and sell embodied transparency may yet redeem a character.

Who Subverted Quiz Shows?

By taking a moral stance against the commercialization of transparency, *Quiz Show* taps a puzzling cultural phenomenon that I can only explain in cognitive terms. Although some of the movie's characters are quick to point out that "this is television after all" and that it's naive to expect that a popular show should limit itself to unscripted emotional responses, these characters are not particularly sympathetic. *Quiz Show* shares at least some of the 1950s viewers' indignation at being cheated out of real competition and real emotions.

Whence comes this indignation? Why should anyone feel angry when they find out that the sweat glistening on the contestants' foreheads isn't a sign of real anxiety but rather a result of applying a moistened handkerchief to create an impression of anxiety? And how can this half-a-century-old anger still resonate with audiences today, enough for the director to center a morality tale around it?

It's possible that we are dealing here with the same phenomenon that I explored in *Why We Read Fiction,* in which I talk about people feeling angry on finding out that stories initially presented to them as real were actually fictional. I suggest in that book (taking as my starting point the work of John Tooby and Leda Cosmides on source-monitoring)[11] that when we know that we are dealing with a fictional narrative, we process information contained in that narrative with a strong source tag pointing to its author, which prevents that information from circulating freely among our cognitive databases and impacting an unforeseeable variety of real-life decisions. Hence adjusting our thinking as a result of learning that a story we thought was true is actually fictional may involve a cognitive cleanup of proportions that we can't even estimate, and it is our intuitive awareness of the need for such an effortful cleanup that underlies our feelings of anger and betrayal.

Here is how this unhappy awareness of the high cognitive cost may manifest itself in specific cultural circumstances. Say the viewers think that Charles Van Doren defeated Herb Stempel fair and square. An endless variety of cultural narratives about social class, ethnicity, and education—narratives whose emotional valence differed widely from one group of viewers to another in 1950s America—swirl around this triumph of a polished privileged WASP over a nerdy middle-class Jew. The emotions provoked by these cultural narratives and their personal interpretations are real, so finding out that the whole thing has been but a performance calls for reevaluation of these emotions and narratives. This is a painful process rendered additionally unpleasant by the fact that one is not sure which personal narratives were impacted in which way. Did I decide not to go to college, after all, because of something that struck me as I was watching Herb Stempel sweating visibly in that little soundproof booth on the stage? And should I trust that intuition about myself now that I know that Stempel was playacting, and that I, obviously, don't know anything about him?

I am channeling a hypothetical young American in the 1950s, but there is no need for such fancy role-playing. Here is writer Gayle Pemberton reminiscing about a very real effect that watching Gloria Lockerman perform on a quiz show in 1955 had on her life. Gloria "was a young black child, like [Gayle], but she could spell anything" and thus "won scads of money on 'The $64,000 Question.'" Gayle's grandmother, usually scathingly critical of everybody's performance, both on TV and in real life, told Gayle that she "ought to try to be like" Gloria.

The remark "shocked" Gayle: "I was . . . thrown into despair. I had done well in school, as well as could be hoped. I was modestly proud of my accomplishments, and given the price I was paying every day—and paying in silence, for I never brought my agonies at school home with me—I didn't need Gloria Lockerman thrown in my face. Gloria Lockerman, like me, on television, spelling. I was perennially an early-round knockout in spelling bees."

Almost forty years later, Pemberton, now a distinguished professor of English at Wesleyan University, muses that "Gloria Lockerman was partially responsible for ruining my life. I might have never ended up teaching literature if it had not been for her."[12] Given the personal impact that Gloria's performance thus had on both adults and children, it's not difficult to imagine how unsettled they would have been by a revelation that her performance and emotions on TV were, too, scripted by the show's producers.

(As a matter of fact, they weren't scripted: several years later, *The $64,000 Question* did become embroiled in the quiz-show scandal, but their problems were of a different kind than those besetting *Twenty-One* and *Tic-Tac Dough*. The show's corporate sponsors wanted to include celebrities as contestants while bumping non-celebrities no matter how well they performed and how much the public liked them. Gloria herself had decided to quit the show early because she didn't want to risk it all on the $64,000 question, following the advice of *her* grandmother.)

This is to say that we want to know the relative truth-value of a context in which we glimpse embodied transparency because it may help us keep track of our own emotions and decisions. Though on some level we don't differentiate between reading the minds of real people and reading those of imagined people—our theory of mind applies itself with a healthy appetite to both—we anticipate engaging in a different kind of personal cognitive management when dealing with real-life as opposed to fictional emotions. The quiz shows promised one kind of cognitive management and delivered another—that was their real "scandal." The movie, taking its cue from the original public reaction, encourages us to think of the shows as having violated some moral norms, but it has a difficult time articulating what these norms actually are, and for good reason. To get at the bottom of how the quiz shows did the public wrong, we have to talk about the high cognitive cost that they imposed on their unsuspecting viewers, but this language is not yet an accepted part of our cultural analysis.

Quiz Show thus reimagines a particular moment in the history of television when what seemed to be a reliable context for embodied transparency—a heated competition between would-be experts in cultural and political trivia—became subverted. The movie portrays television producers and corporate sponsors as villainous agents behind this subversion, but what we have learned about embodied transparency so far ought to make us less eager to assign blame.

For this is the way of embodied transparency. It wanes as soon as it starts to blossom. By the time you finish your heartfelt scream at the sight of the mouse, you might've already been faking it for two seconds. A quiz show is no more immune to its participants' performance of body language than was an eighteenth-century sentimental novel or is twenty-first-century reality-television programming. Blame it on the cognitive phenomenon we are dealing with (our perception of the body as simultaneously the best and the most suspect source of information about the mind), and accept that the naked emotional face that we were prevented from seeing would have been the face of a performer.

SIX

{
In which the makers of *The Office* reveal their recipe for making us cringe / Hubert Humphrey and John F. Kennedy don't care how they appear onscreen / film directors hunt for body language that can't be faked / and Andy Kaufman talks of pulling off his own death.
}

Mockumentaries, Photography, and Stand-Up Comedy
Upping the Agony

Cringefest

It's another day at *The Office*—a British mock documentary about the "boss from hell," David Brent (Ricky Gervais). A regional manager of a paper company, David is about to interview two candidates for the position of secretary. This must be the worst time to hire new people—the branch is about to be downsized, and everybody is anxious about losing their jobs. But, as David pompously explains to the camera that follows him and his coworkers around the office, he "needs a secretary," so he is getting one.[1]

The "lucky contestants" (as David calls them) are a man and a woman in their late twenties. David immediately focuses on the attractive woman. "She'll brighten up the place, won't she?"—he observes to Dawn (Lucy Davis), the office receptionist, in the full hearing of both candidates, then catches himself and adds, "if she gets the job." "So will you"—he turns to the man—"because you are both equal. No foregone conclusion. Based on interview and merit. It is up to me ultimately. But good luck. You'll do well to impress me."

Used to David's ways yet painfully ashamed of his antics in front of these two strangers, Dawn stands next to him silently. She knows that she is being doubly observed—both by the interviewees and by the documentary film crew. She can't afford to be caught on film openly rolling her eyes and expressing scorn at her boss's behavior. So she keeps smoothing her hair and checking her nails, while avoiding any eye contact with the relentless camera (fig. 20).

And it seems for a moment that Dawn will be able to escape the camera, busying herself with something else. Enraptured by the sexy interviewee, David decides to take a picture of her—"Just for um . . . just for er . . . just for the files"—and sends Dawn down for a Polaroid. But when she brings it, he grabs it himself—"I'll do it. I'll do it"—and moves closer to the female candidate. "Let's get that lovely smile on. . . . That's nice. Oh, the hair all . . . lovely. That looks lovely. Lovely blue eyes. OK. Big smile. That was lovely. . . . Look at that."

As David is waiting for the picture to develop, he remembers the other candidate. "Do one of you as well," he throws off and clicks the shutter in the young man's general direction. "We are [interviewing] Stuart first," Dawn reminds him in a hopeless attempt to maintain some semblance of decorum. "Let's get him out of the way," David agrees volubly. "Follow me." The shot ends with Stuart getting up to follow David to his office and Dawn closing the Polaroid camera, too embarrassed to look at either candidate.

The Office feasts on such awkward moments. The documentary format allows the film crew to focus pitilessly on people's faces just when they would rather not be seen, encouraging the kind of staring that would be considered rude in real life. As Gervais, who codirected the miniseries with Stephen Merchant, puts it, one "advantage" of having the ever-watchful camera in *The Office* is "that it would wind up the agony."[2] And although we may share some of that agony as we watch Dawn cringe and squirm, we remain glued to our screens.[3] Dawn is transparent, she knows she is transparent, she tries not to be, she can't help it, and we get to see it all.

Witness the first rule of transparency: *The Office* cultivates contrasts. Gervais and Merchant constantly prod us to gauge one character's trans-

FIGURE 20. Dawn (Lucy Davis) and David (Ricky Gervais) in Gervais and Stephen Merchant's *The Office* (BBC, 2001–3).

parency as contrasted to that of other characters or to her own transparency a moment ago. For example, when David makes a fool of himself in front of Dawn and the two job candidates, both interviewees must be unpleasantly surprised by his behavior. Still, the scene is shot so as to emphasize the contrast between what we may infer about their respective feelings and what we may infer about Dawn's feelings. We can see that Dawn goes through the agony of embarrassment *and* tries to conceal her disapproval of David's actions. Because we can imagine how much mental energy this must take, we feel that it is highly unlikely that Dawn is capable of thinking of anything else at this moment (say, about her fiancé, Lee; her coworker, Tim; or the book that she was reading earlier). Just now Dawn is strikingly transparent.

Not so Karen and Stuart, the interviewees. Karen smiles a lot, and at one point she also shoots a curious glance at Dawn. We may infer that she is taken aback by David's lack of professionalism but determined to get through the interview. Other than making this tentative inference, however, we have no way of knowing what goes through her head as she listens to David's harangue.

Stuart is even less transparent than Karen. We may guess that he is disappointed and angry and that he has already given up on the job and is now only keeping up the appearance of polite interest because the camera is present. These speculations, however, draw more on the context of the scene than on his body language. For he wears a small, noncommittal smile throughout and otherwise betrays no emotions.

What about the rule of transience? Did the makers of *The Office* do something special to keep the instances of embodied transparency, such as the one above, brief?

It turns out that they did, even if they don't think about it in these terms. In their commentary to *The Office* Merchant and Gervais explain that they wanted to avoid the feel of "situation comedy," so that their material would remain jumpy and raw. To achieve that, they made a point of cutting abruptly from one scene to another. Note, however, one important side effect of this editing technique: it ensures that when characters are forced into a state of embodied transparency, we typically don't stay with them long enough for them to seize control of the situation and start performing their feelings.

This really is an effective narrative trick. *The Office* doesn't dwell on the same person for a long time, yet it leaves us under a strong illusion that it does, by making us aware that we stare at the protagonists when they are embarrassed or making fools of themselves for longer than would be polite in real life. So we are made to feel (as we do with Dawn) that we have seen too much when, in fact, we have seen just enough to be convinced that the protagonists were caught off guard and didn't have time to rally their spirits and put on a suitable performance.

Death or Fiction: Cinéma Vérité's Choice?

With its treatment of embodied transparency the mock documentary occupies a peculiar position in relation to its grandparent genre of regular documentary and its parent genre, cinéma vérité.[4] In its ambitions it takes after the parent, in the execution of these ambitions after the grandparent. The result is pure fiction: embodied transparency galore, and all of it fake. The documentary roots are important however. By looking at them we can reconstruct the history of a genre—in this case, of cinéma vérité—struggling to maintain its claim to portraying "real" emotions in a culture suspicious of such claims.

Cinéma vérité thrived on the spontaneous show of emotion—indeed actively looked for contexts that would allow for maximum transparency. As Hope Ryden, who wrote, directed, and produced documentary films from 1961 through 1987 and was part of the Drew Associate team that developed cinéma vérité/cinema direct in the early 1960s,[5] puts it, "What we were doing was finding an upcoming event in which some character would have a great stake. At that moment they would either win or lose. And it didn't matter whether they win or lose. What mattered [was] that they cared a whole lot about what they were doing."[6]

And what this caring "a whole lot" meant for the movie was that no matter what the characters might do on camera during their make-or-break moment, the viewer knew what they were really feeling. So if they showed emotions, such as anxiety, happiness, or disappointment, those emotions could be counted on as being authentic. And if the characters didn't show any emotion, they would still be transparent, because the viewer knew that they were trying hard to conceal their feelings.

In the words of Robert Drew, who directed the documentary *Primary* (1960), featuring presidential candidates Hubert Humphrey and John F. Kennedy during the Wisconsin primary elections: "The idea of capturing human emotion spontaneously as it happens was the key idea that made *Primary* work, that made all of our films work, and that is making cinéma vérité work today in many ways across the spectrum of television.[7] As historians of documentary film Jack C. Ellis and Betsy A. McLane put it:

> [In] *Primary*, Humphrey and Kennedy were much more concerned with winning an election than with how they would appear on screen. . . . *Mooney v. Fowle* (1961, aka *Football*) builds up the climaxes with a high school football game in Miami, Florida, between two rival teams. It concentrates on the players, coaches, immediate families—those most completely preoccupied with this contest. *The Chair* (1962) centers on the efforts of a Chicago attorney, Donald Page Moore, to obtain a stay of execution for his client, Paul Crump, five days before it is scheduled to take place. *Jane* (1962) concerns

Jane Fonda in the production of a play, from the rehearsal period through the negative reviews following its Broadway opening and the decision to close it.

Thus when in *The Chair* "the attorney breaks into tears and expresses his incredulity after he receives a phone call from a stranger offering support for him in his efforts to save his client's life,"[8] we cannot possibly doubt that we see the man at his most transparent, that his body language provides direct access to his mind.

The creators of cinéma vérité seemed to have achieved the impossible: a representation of real feelings—not a carefully arranged tableau of involuntary emotions (as in paintings), not a description of spontaneous body language of people who never existed (fiction), not a performance of unmediated emotional responses by professional actors (theater and movies) but actual transparency exhibited by actual people yet carefully observed and recorded by the camera.

But, as we've seen, once a culture is aware of a seemingly reliable representational context for embodied transparency, that context is rendered suspect. Theater becomes a place where spectators fake their exalted emotional reactions, horse races—where they fake lively interest in horses. We learn to read emotional fakery into a show of feelings that yesterday we took at face value.

The history of cinéma vérité can be viewed as an attempt on the part of documentary filmmakers to stay ahead of this particular learning curve. They knew from the beginning that their claims of direct access were fragile. Ideally, their subjects should "reveal what they really felt and were like when unselfconsciously relaxed or deeply involved in some activity."[9] In reality, however, the presence of the camera must still have influenced emotional responses of even the most unselfconscious subjects. As Jean Rouch, the director who originated the term *cinéma vérité*, saw it, "the camera acts as a stimulant. It causes people to think about themselves as they may not be used to doing and to express their feelings in ways they ordinarily would not."[10] Performance worms its way into the vérité (however much vérité there was to begin with).

It was only a question of time, then, before viewers would grow skeptical and start distrusting the whole genre. And if we assume that directors wanted to anticipate this reaction, then we can look at certain films made between the late 1960s and now as their attempts to find new, reliable contexts for embodied transparency.[11]

Among such attempts we should count documentaries that depict people killed during rock concerts (e.g., *Gimme Shelter,* 1970), victims of war in Vietnam (*Hearts and Minds,* 1974), people dying of AIDS (*The Broadcast Tapes of Dr. Peter,* 1994), and people literally growing up on camera (Michael Apted's "Up Series" of 1970, 1977, 1984, 1991, 1998, and 2005). All of these films depict physiological processes that cannot be faked; hence, they elicit a very strong feeling that, at least on some level, these bodies provide direct access to the minds. For instance, Apted's subjects, whom he filmed every seven years, can say whatever they want about themselves, but we find it particularly gratifying when we can correlate what we hear or see (e.g., voice pitch, posture, makeup) with the fact that they are now seven years older than they were in the last installment.

It's striking how many of these films portray death: the ultimate instance of embodied transparency. Of course, with the exception of certain unique circumstances, such as those created by the epidemic of AIDS in the 1980s and 1990s,[12] the directors did not and could not set out expecting to film *that* kind of transparency. (For instance, Albert and David Maysles and Charlotte Zwerin could not know that a person would be killed on camera during the performance of the Rolling Stones in their film *Gimme Shelter.*)[13] It is interesting, then, that the 1970s saw a number of productions *appearing* to be documentaries shot in cinéma vérité style, such as *No Lies* (1973), a "staged film about rape," and *Rushes* (1979), a staged film about suicide.[14] In the wake of the original cinéma vérité of the 1960s these fake documentaries demonstrated the genre's intuitive awareness that "deliberate and extraordinary measures"[15] might eventually be needed to sustain its claims to direct access. And they were right to the extent to which, in the years to come, the real cinéma vérité did turn to death and other unfakeable physiological experiences, such as growing up.

But where is a genre to go in order to retain its claim to emotional authenticity after it has reached the point of documenting unfakeable physiological experiences? Crossing over into the realm of fiction seems almost inevitable. So the mock documentary is an offshoot of cinéma vérité's obsession with direct access to emotions crossbred with techniques for creating illusions of such access that are actually alien to cinéma vérité.

For instance, while cinéma vérité merely *hoped* to catch at least some moments of embodied transparency and looked for situations likely to produce such moments, a mockumentary actively *transforms* every situation into an occasion for embodied transparency. As one brief illustration of how this transformation happens, consider Gervais's commentary on the "talking heads" of *The Office*. He refers to the scenes in which the protagonists are interviewed individually and thus can presumably put on any attitude or personality:

> I really love the talking heads in the show. Because we shot it like a documentary, we couldn't do things people wouldn't do in front of the camera. They can't shut the door and take a line of coke. Or they can't blurt out things they're thinking.
>
> But, ironically, when they're by themselves and they're just being filmed, they're a little bit more honest. People do let their guard down because it's flattering. When a camera's pointed at someone, they think, "This is my chance, this is my platform. I can tell the world my all great philosophies on life" [*sic*]. And of course, they open their mouth and they blow it and they can't take it back.

Note what just happened here. The least likely moment for revealing one's true thoughts is turned into its opposite. It's one thing to be *caught* on camera as Dawn is in the scene with David and the two job candidates; it's quite another to be invited into a separate room to be filmed during a formal interview. In the latter case you have all the opportunities in the world to prepare well and to put forth your better self. After all, many early cinéma vérité directors prided themselves on *never* interviewing their subjects. Interviews were the mainstay of stodgy traditional documentaries—they spelled out performance. But in *The Office* this context-for-performance par excellence becomes yet another context for transparency. The protagonists "blow it and can't take it back," and they know that their audience will be able to tell that they are not happy about it.

On this and other occasions, thrusting the camera into people's faces is a technique of old documentaries that cinéma vérité abhorred but that is crucial to *The Office*. Ironically, *The Office* needs it in order to capture "real" emotions—which is what cinéma vérité wanted to capture, too, except that cinéma vérité hoped that its subjects would forget about the presence of the camera, while *The Office* makes sure that nobody ever forgets about it. This is why the signature "real" emotion that *The Office* captures—acute embarrassment prompted by the awareness of being watched in a socially awkward situation—would be alien to the world of cinéma vérité. Not to mention that this embarrassment is fictitious, anyway . . .

Fake self-consciousness—is this what cinéma vérité has come to with *The Office*? Tangled are the ways of a genre that pursues embodied transparency in a culture that seems to consume formerly reliable contexts for transparency faster and faster.

Death and Photography

Embodied transparency in photography deserves a separate chapter, with subsections on candid photography and street photography, on Henri Cartier-Bresson and Arthur Fellig (Weegee), on Susan Sontag and Roland Barthes. I don't deal with photography in this book, though, and bring it up now only briefly as a point of comparison for my discussion of cinéma vérité's gradual turn to unfakeable physiological experiences in the 1970s and 1980s.

Generally, photographers seeking to capture unmediated emotions face even greater challenges than do documentary filmmakers. Unlike the latter, they don't have the advantage of a narrative that leads up to the presumed moment of direct access and thus assures viewers (at least to a point) that the emotions they see weren't merely performed for the camera. This may explain why photography often turns to subjects that come with their own, built-in, so to speak, embodied transparency, such as infants and young children (see fig. 21 as an illustration of the ease of arranging a shot that captures a child's "true" feelings, such as, in this case, his absorption in a puppet); attendees of engrossing performances and athletic events (such as Weegee's "Teen Age Audience"); and people whose range of likely thoughts and feelings has been drastically narrowed by famine, impending execution, and other life-threatening circumstances.

FIGURE 21. A child with a puppet. Photo courtesy Brian Connors Manke.

This means that when photographers explicitly set out to record "true" mental states, they can expect, unhappily, one of the two following reactions from their audience. On the one hand, spectators are always eager and ready to interpret photographic instances of direct access as still somewhat fake, that is, constructed, staged—*importing* emotions into a context instead of merely recording them. The reason for this suspicious attitude is the same as I discussed earlier: because we evolved to perceive the body as the best yet most unreliable source of information about mental states, we remain skeptical about any context that promises us guaranteed access to mental states.

On the other hand, if we do have a context backed up by the ultimate guarantee—that is, if a photograph depicts a life-and-death situation, which often means that the subject will die as soon as the picture is taken or is already dead—and the photographer, therefore, cannot be suspected of staging a particular context or emotion, she can be accused of not acting to prevent the death that she was recording. She won't be prosecuted, but the impression of cruelty and collusion will linger.

In other words, embodied transparency in photography is often either not convincing enough—unless it's too simple, as in pictures of babies—or morally indefensible. It's hard to get a shot of a "real" feeling in a socially complex and non-life-threatening situation.

Death and Stand-Up Comedy

To abandon the screen and other flat surfaces for the stage, one can argue that the fragility of any established cultural context for embodied transparency is what drove the career of the brilliant American stand-up comedian Andy Kaufman. Kaufman built his onstage presence around transparency, exploiting various social contexts in which a person *must* inadvertently show his true feelings: for example, pain and disappointment after being heckled by the audience; acute embarrassment when things don't work out the way they are supposed to; quiet happiness after a profound spiritual experience changes one's life and allows an almost dead career to take off again. All of these contexts were faked, of course. In fact, in a paradoxical act of reversing the roles of the observer and observed, Kaufman must have enjoyed the transparency of his spectators as he watched them growing uncomfortable, impatient, or angry during his stage antics. (Think: if Dawn from *The Office* could see us squirm in front of the television as we watch her being painfully embarrassed, who'd be more transparent then—she or we?)

As time went by, however, audiences wised up to Kaufman's strategy of provoking them. According to one biographer, by the early 1980s "fewer and fewer were fooled and fewer still would care to be."[16] In response Kaufman felt compelled to come up with more and more outrageous scenarios to convince people that *this time* they really could see into his soul. And, inevitably, after a while, nothing short of being physically hurt or dying would seem to have sufficed to prove that his body provided at least some access to his true feelings.

Kaufman's friend and collaborator, Bob Zmuda, remembers Kaufman saying, "with boyish enthusiasm," after reading one "particularly vicious" review of his work: "This one says I've gone as far as I can go, short of actually killing myself on stage. What do you think?"[17] In 1982 Kaufman confided to the producer of *Saturday Night Live*, Bob Tischler, "You know, the hoax I'd really like to pull off is my death. But I'm afraid of doing it—because when I do these things, I do them for real, and so I wouldn't even be able to tell my parents. And I wouldn't want to hurt them." At that time he was already talking to his friends and relatives about wanting to pretend "to have cancer or something."[18] This was after the Jerry Lawler episode—when Kaufman convincingly pretended to be seriously injured by the angry professional wrestler—but before he was actually diagnosed with lung cancer, in December 1983.

Predictably, many people thought that he faked his subsequent death in May 1984 and staged his own funeral. Once doubt creeps in, even the ultimate embodied transparency can be compromised.

SEVEN

In which emotions build up delightfully / a bachelor turns down a girl (who promptly blames herself) / individuals who don't like reality TV show their true faces / a TV executive claims that people like being humiliated / *American Idol* is mentioned all too briefly / and a lot of mental energy is spent on strangers.

Reality TV
Humiliation in Real Time

Overwhelmed by Emotions

N ow what about reality shows? Clearly, embodied transparency is a huge factor in their appeal. That said, the term *reality show* covers such a variety of formats and approaches that we should be careful about claiming that *all* they do is cultivate moments when people's bodies betray their feelings. This is in direct contrast to my discussion of such shows as *The Office*. Because mock documentaries represent only a small segment of television programming, I feel justified in saying that they literally exist to cultivate moments of embodied transparency. I would not, however, make the same overarching claim about the rapidly expanding universe of reality shows. Mine is a more modest claim: the concept of embodied transparency and the cognitive framework that underlies it offer us useful tools for understanding certain recurring features of these shows.

One such feature involves a buildup toward a predictable emotional response. That is, some shows are so arranged that as we watch them, we get closer and closer to one pivotal moment at the end of the episode when a participant will feel a particular emotion. There is no surprise in it for us: we know exactly what that emotion will be. For instance, in *Coming Home*, a series featuring unexpected military reunions, we know that spouses, parents, and children of returning soldiers will be overcome by joy and relief at the sight of their loved ones. Just so in *Extreme Makeover: Home Edition*, we know that the owners of a rundown house will be overwhelmed with delight and gratitude when they first see the dream home built for them by the remodeling crew. Our knowledge, however, does not dampen our eager anticipation and subsequent pleasure as we watch the participants finally experiencing that emotion. In fact, our pleasure *is* the pleasure of knowing for sure—and having imagined all along—what the participants *really* feel as they cry, or laugh, or just stand there dumbstruck.

Now think of trailers used to promote reality shows. Sometimes we see the face of a participant going through a strong emotion, like crying. Given the context of this series—say, if it's a dating show, such as *Bachelor*, in which several women compete for the same man—we can infer that the participant has not been chosen by the desirable bachelor and is now bitterly disappointed and dissatisfied with herself. Note how we are being lured in by this glimpse of embodied transparency—by the promise of perfect access to the participant's feelings.

A closely related trailer strategy is to show a participant at the moment when his body language is relatively muted while the voice-over explains that we are witnessing a very particular social situation—for example, that somebody has committed a social faux pas, which has put our protagonist in a strikingly awkward position. We know that were *we* to find ourselves in this position, we would experience very strong feelings of surprise, humiliation, or anger, So we infer that he is, in fact, surprised, humiliated, or angered but, because of social conventions, cannot express it just now. So the participant is doubly transparent to us: not only do we know the gamut of negative feelings he is experiencing, but we also know that he labors mightily to refrain from displaying them.[1] Again, this is a bait that we may find hard to resist (i.e., a glimpse of direct access to a complex mental state), which is the reason that this moment is chosen to advertise the episode.

Note, by the way, that by saying *we* so often I don't want to imply that reality shows are irresistible for everybody. Some people don't watch any of them. What I mean is that, while like any form of entertainment, they appeal more to some people than to others, they build their appeal around techniques that tap human cognitive universals, such as our ability, need, and desire to read minds in social contexts. The same argument can be made about reading novels, admittedly a more respectable cultural pastime: novels are built around our theory of mind, but not everybody is an avid novel reader.[2]

Humiliation: Pleasure or a Means to an End?

Humiliation comes up frequently in public discussions of reality shows. In 2003 the Museum of Television and Radio in Los Angeles sponsored a seminar titled "The Past, Present, and Future of Reality Television." Humiliation was the subject of one of the first questions that the moderator, Barbara Dixon, asked the panel of reality television producers and broadcast executives. As she put it, "Do we take pleasure in seeing people humiliated? What is that about? What goes into that formula that makes it seem so popular?"

In response, producer Scott A. Stone noted that contemporary reality shows' precursors, such as *Queen for a Day* and *Candid Camera,* were about humiliation as well. According to Stone, this is "just part of the audience for television: they want to see humiliation." Mike Darnell, executive vice president of alternative programming for Fox TV, disagreed, observing that participants themselves do not feel put down: "the more humiliating the program seems to be, the more people show up for auditions." In Darnell's view it's not humiliation that attracts the audience; it's "a voyeuristic desire to see people go through emotional struggles."

Stone's and Darnell's views are actually not as different as they seem once we consider embodied transparency as something that producers of the show want to achieve. (Not, of course, that they think about it in these terms.) That is, if the goal of the producers is to put participants into a situation in which their thoughts are obvious to the audience even as they struggle to conceal them and maintain their cool, then creating a context in which participants are humiliated is one obvious and winning strategy. So we can agree with Stone when he insists that television audiences want to see humiliation, but we can also say that it's not humiliation per se that they are after but access to participants' thoughts and feelings. Humiliation is a direct path to that access, a means to the end.

And that's why what Darnell is saying is perfectly compatible with Stone's view. We want to see "people go through emotional struggles" because this gives us access to their feelings, and being publicly humiliated necessarily involves an emotional struggle. To repeat, humiliation is not the *only* strategy for making people struggle with themselves, thus becoming transparent to the audience, but it happens to be reliable and effective, so it's used a lot. Think of my earlier discussion of restraint. Just as there is nothing irresistibly attractive about restraint when it comes to movies, there is nothing intrinsically fascinating about humiliation when it comes to reality shows—both work to bring about transparency, and that's why they appeal to directors and producers.

Moreover, there is no reason that the same program cannot combine one particular strategy with another. Contest shows, such as *American Idol,* use humiliation *and* a buildup of predictable emotions. They humiliate their participants early on, during the auditions, and they also build up to the exhibition of strong emotions in the finale, when the winners rejoice and the runners-up break down in tears or struggle to hide their disappointment (as we all along expected they would!).

Why Watch Reruns?

Note, by the way, how the claim that I am making here is different from a closely related claim, often made about reality shows, which is that we like watching them because we like watching people's emotions. On the one hand, I agree with this completely and can further say that we spend time, energy, and money to put ourselves in situations in which we can do just that (by renting movies, by going to theaters and museums, etc.). On the other hand, unless we bring in research on theory of mind, we have no answer to a simple follow-up question: *why* do we like watching people's emotions? It's only when we start thinking of mind reading as our most crucial and constant preoccupation (though not consciously so) as a social species that we can say that we like watching displays of emotion because they promise access to people's thoughts, feelings, and intentions, and we evolved to value such access tremendously.

Moreover, reality shows seem to offer their viewers a form of mind reading not available in movies and plays because they focus on emotional reactions of ordinary people and not professional actors. The assumption here is that participants of such shows are not good at faking or concealing their feelings. In fact, they might be quite bad at it (an impression lovingly cultivated by strategic close-ups of their faces). Putting people who are not trained to perform their emotions in situations that surprise, unsettle, or humiliate them thus seems to guarantee that what we will get is unmediated access to their true mental states, which is a valuable and difficult-to-come-by social commodity.

A completely useless commodity in this particular case, since we are not engaged in any direct social interaction with the participants in the shows. Our mind-reading adaptations, however, do not quite "get" this disconnect from reality. They evolved in an environment that didn't have television sets, which means that people whose intentions and feelings we had to figure out were, in fact, our (potential) mates, friends, rivals, and enemies. Successes and failures of our mind reading back in the Pleistocene had immediate and serious repercussions. This is why today we not only hang upon every twitch of a brow of an anonymous passerby set up by the cast of *The Joe Schmo Show*, but we also do it repeatedly, via reruns. As Sheryl Longin, a columnist at *Pajamas Media*, writes in her essay "Confessions of a Reality Junkie,"

> I asked my daughter why she liked to watch these same episodes
> over and over. She stated the obvious fact that real people are more
> interesting to watch. True enough, voyeurism is humanity's guilty
> pleasure. But there is nothing unpredictable or shocking about the be-
> havior exhibited by the participants in these shows. In fact, the oppo-
> site is true. They are fun to watch precisely because they are familiar
> character types we have all encountered in our daily lives. And any
> rare unexpected behaviors they do exhibit are no longer surprising on
> second viewing. Yet the fascination remains.[3]

The fascination is bound to remain because as a mind-reading species we are uniquely vulnerable to representations that seem to give us direct access to people's minds; because professionals who create such representations use a variety of editorial techniques and scripts to intensify this illusion of access; and because there was no concept of reruns in the Pleistocene. That is, there simply had been no circumstances in which people would perfectly replicate their emotional displays, which might have conditioned our mind-reading adaptations to treat such displays as meaningless the second time around.[4] Hence, when faced with reruns, these adaptations continue to enthusiastically process participants' body language, picking up (or imagining) new cues and nuances, and making us feel that we engage in a social activity that is important and meaningful.

Of course, we may feel guilty afterward, when the faces of our "social partners," which we've been attending to so intently for the last hour, are replaced by commercials and we almost wake up to the fact that we just wasted a lot of mental effort on total strangers. Come next episode, however, we are back: back to knowing exactly what goes through their minds—brilliant social players that we are made to feel by the show's producers. What chance do we have, really, we with our prehistoric mind-reading adaptations, against an industry that is daily figuring out new ways to make these adaptations hum with pleasure?

EIGHT

In which Colonel Pickering can hear Higgins in *My Fair Lady*, but Emile can't hear Nellie in *South Pacific* / Sky Masterson slows down time in *Guys and Dolls* / Audrey Hepburn walks in on one eleven-o'clock number, and Rosalind Russell takes her turn for another / "Mr. Cellophane," in *Chicago*, is shown not to be that transparent / Sondheim turns it backward in *Merrily We Roll Along* and inside out in *Sunday in the Park with George* / a student and a girl talk up a storm in a thirteenth-century Chinese opera / Evan Rachel Wood complains about singing and acting at the same time / and the author timidly puts forth a suggestion.

Musicals
(Particularly around 11 pm)

Sing Your Heart Out

The expression "sing your heart out" is a cliché, but it captures perfectly an intuitive mind-reading expectation we have about singing in stage and screen musicals: singing gives us direct access to characters' hearts, revealing their secret thoughts, feelings, and desires.[1]

Of course, not every instance of belting out a song provides a direct pathway to hidden feelings. That mostly happens when the scene is arranged in such a way that other characters—even those standing right next to the character who is singing—cannot hear her. To clarify this point, remember that there are three distinct ways of introducing a song into a play. According to theater historian Scott McMillin, one way is to have characters "deliberately perform numbers for other characters." There is a whole tradition of "backstage" musicals—stories about performers who put on musicals: "Since the characters are show people whose job is song and dance, much of the singing and dancing is called for by the book.

Show Boat is about entertainers, so they sing and dance. . . . *Phantom of the Opera* uses an opera-within-the-musical device—the plot occurs at the Paris Opera where various operas, including one by the Phantom himself, are being rehearsed and staged. *Follies* takes place at a reunion of former Follies showgirls, who perform some of their old numbers for one another."[2]

Another way to introduce a song into a musical is to have a character sing as he speaks to other characters. For instance, when in the opening scene of *My Fair Lady* (1964), Higgins sings "Why Can't the English [Teach Their Children How to Speak]," he directs his harangue at Colonel Pickering, Eliza, and everybody within earshot in Covent Garden. And when toward the end of the musical, he sings, "Why can't a woman be more like a man?" he is addressing first Pickering and then his own housekeeper.

In both cases—when a character deliberately performs a musical number and when he sings while talking to others—other people onstage hear him.[3] In contrast, a character can burst into a song that seems to "come from out of the blue" and that nobody else in the world of the play can hear.[4] Such songs are essentially thoughts: as long as the character is singing, we know what he is thinking.

As an example of this kind of embodied transparency in a musical, consider the 1958 film version of Rodgers and Hammerstein's *South Pacific*. To make it abundantly clear that what we get is neither a song-as-performance nor a song-as-conversation but a song as the direct and exhaustive report of the protagonists' thoughts and feelings, they are shown at one point "singing" with their mouths shut. This is the scene in which Nellie Forbush is visiting Emile de Becque's plantation and there is a pause in their conversation as he goes toward the table to pour the wine while she gazes at the ocean. Neither says a word, but we hear their respective "thoughts" sung as voice-overs:

NELLIE. Wonder how I'd feel living on a hillside,
Looking at an ocean, beautiful and still.
EMILE. This is what I need. This is what I longed for:
Someone young and smiling, climbing up my hill.
NELLIE. We are not alike. Probably I'd bore him.
He is a cultured Frenchman. I am a little hick.

EMILE. Younger men than I, officers and doctors,
Probably pursue her. She could have her pick.
NELLIE. Wonder how I feel jittery and jumpy.
I am like a schoolgirl waiting for the dance.
EMILE. Should I ask her now? I am like a schoolboy.
What will be her answer? Do I have a chance?

This is a delightful moment for many reasons; it doesn't hurt, for example, that the actors, Mitzi Gaynor and Rossano Brazzi, are attractive and charismatic. A crucial factor, however, that contributes to our feeling of pleasure as we watch it is the pattern of mind reading that the scene imposes on us. When Nellie sings in the voice-over, "He is a cultured Frenchman. I am a little hick" (fig. 22), and we see her first turn ever so slightly in the direction of Emile and then turn away with a barely perceptible frown at "hick," our mind-reading adaptations must be most happily occupied by matching her body language with her thoughts. This matching continues as Emile pauses with a concerned look, a wine bottle

FIGURE 22. Nellie (Mitzi Gaynor) thinks in song in Rodgers and Hammerstein's *South Pacific* (South Pacific Enterprises, 1958).

FIGURE 23. Emile (Rossano Brazzi) thinks in song in Rodgers and Hammerstein's *South Pacific* (South Pacific Enterprises, 1958).

in one hand and a glass in another (fig. 23), thinking of the younger men who "probably pursue" Nellie. The matching continues still as Nellie half-smiles, comparing herself to a romantic schoolgirl, and as Emile pauses once more and gives another tiny frown as he compares himself to a schoolboy.

We match, we correlate, we infer. That is, we engage in an intensive cognitive workout. Not only does this experience simulate everyday mind reading, but it also provides us with an improbably perfect correspondence between body language and thought. We infer that Nellie frowns because she is thinking that she is a little hick, and *that's exactly why she frowns.* Wow—we are good!

If this seems too obvious, consider that in real life when you think that you know why you just frowned and will readily explain it to anybody who asks, there is a possibility that something else was on your mind: some vague complicating nuance that was too tangled or boring to go into when another perfectly plausible explanation was at hand. Or, just as likely, there might have been something else behind it, something you yourself were not aware of at the time. In other words it's not at all clear that we can reliably talk about knowing our own "real" mental states—much less those of other people (again, life-threatening situations excepted).[5]

This doesn't mean that all human communication is travesty—we manage to get through the day, and even accomplish something now and then, without making our inner thought processes transparent to ourselves and to others. This does mean, however, that a scene from a musical, such as the one above, can create an illusion of complete and exhaustive correspondence between the mind and the body in a complex social situation that has no analogue in real-life communication and must therefore feel like an unusual treat to our theory of mind.

Once again, this scene temporarily turns us into improbably brilliant social players. For even though Emile's and Nellie's body language is far from neutral, it is not very expressive either. In fact, it is self-consciously subdued. No other character present on the scene would've guessed from their pauses, half-turns, half-smiles, and half-frowns what is going on in their heads. Moreover, given that both of them feel quite insecure, they must be making an extra effort to appear casual and not show just how far they have gone in their hopes. They even stand with their backs to each other, perhaps to better conceal their feelings. Yet we see through their attempts at concealment. We know exactly what lies beneath. Just now our social competence is superlative and effortless.

It fades away slowly. Traces of superior social discernment linger after the song is over, influencing our interpretation of what happens next. When Nellie and Emile resume their conversation and Nellie observes, "I bet you read a lot," we read anxiety into her tone—certainly more anxiety than she lets on—because we know that a moment ago she thought that as a "cultured Frenchman" Emile would soon get bored around an uncultured "little hick." Emile, of course, doesn't even notice her comment about his reading prowess, and we suspect why: he must still be preoccupied with all those images of younger suitors pursuing Nellie. Or perhaps he is thinking of what he is about to say to her? We are moving back to the realm of informed guesses and speculations. The perfect transparency is over.

This brings us to the peculiar treatment of time in musicals. Characters seem to remain transparent as long as their song lasts, which can be three minutes or even longer. Does this violate our "rule of transience," according to which moments of embodied transparency are supposed to be brief?

As a matter of fact, it doesn't. The musical's time moves according to laws of its own. Songs-as-performances and songs-as-conversations unfold mostly in real time, but for songs-as-thoughts, time can stand still or slow down. One technique used by directors to make it seem that the song-as-thought (i.e., the moment of transparency) lasts less than the time it takes to actually sing it, is a freeze-frame. When in *Guys and Dolls* Sky Masterson (Marlon Brando) sings "Luck Be a Lady Tonight" (fig. 24), the actual song lasts three minutes. The scene is done in such a way, however, that it seems that for other gamblers, who are impatiently waiting for Sky to roll his dice, less than three minutes elapses: perhaps half as much of that time. Their body language slows down—even freezes to some extent—while we are inside Sky's mind, listening to his plea to Lady Luck.[6]

Because no other characters are present in the scene from *South Pacific* when Nellie and Emile sing out their thoughts, the directors don't have to use any special techniques to show that time has slowed down. We assume that their songs take as much time as their thoughts do. Or as little time—for when it comes to thoughts (and dreams) we often have no way of knowing how long they actually last.

"To be or not to be" (an Aside)

Here is another important exception to our rule of transience: the rule applies mainly when others are present. When a character is alone (or, as in the case of *South Pacific*, with another equally transparent character), they can be transparent for a great length of time. This is how traditional theatrical soliloquies work, too. "To be or not to be" presupposes a protagonist standing alone on the stage and remaining transparent as long as the soliloquy lasts—for what we are hearing are his thoughts.

Of course, not all directors opt for an unambiguous transparency lasting this long, and the concept "alone" can be stretched and manipulated to complicate our impression of direct access. Thus, in Laurence Olivier's *Hamlet* Olivier recites most of the monologue out loud, except for the part starting from "to die: to sleep;/No more" and until "perchance to dream," which comes in a voice-over. This creates a peculiar effect of a hierarchy of access. The thoughts that we hear in the voice-over, when Hamlet's lips are not moving and his eyes are closed, are somehow more

FIGURE 24. The freeze-frame shot behind Sky Masterson (Marlon Brando) in Joseph L. Mankiewicz's *Guys and Dolls* (Samuel Goldwyn, 1955).

"inner," more "true" than the rest, because nobody can possibly hear them. This is a bit ironic because nobody should be able to hear the rest of the monologue either—Hamlet recites it on a solitary rampart by a raging sea—yet by closing his mouth and eyes, Olivier intensifies the impression of being alone: of having no audience and thus no social incentive to perform his feelings for others.

Every version of *Hamlet* that I've seen answers somewhat differently the question of how much direct access to the protagonist's feelings this monologue should provide. If in Olivier's version the "voice-over" parts of the soliloquy create an illusion of deeper transparency than the other parts, in Kenneth Branagh's version, transparency feels compromised throughout because Hamlet delivers his monologue standing in front of a two-way mirror, almost nose-to-nose with Claudius and Polonius, who are watching him from the other side. With David Tennant's Hamlet, transparency is much less ambiguous. True, the room in which he is speaking is bugged, but Hamlet seems to have found a corner inaccessible to the secret cameras. As Olivier before him, Tennant closes his eyes on "to die: to sleep;/No more" (though his mouth keeps moving—in contrast to Olivier's voice-over), thus intensifying our impression that we are reaching deeper into the character's thoughts. A character who speaks with closed eyes in an empty room is not performing his feelings for others.

Or so this convention used to imply. Perhaps you can already think of some play or movie in which it's been subverted.

Eleven-O'clock Transparency

This is Henry Higgins singing in the final scene of *My Fair Lady:*

> I've grown accustomed to her face.
> I've grown accustomed to the trace
> Of something in the air.
> Accustomed to her face.

Now that Eliza Doolittle has left him, he realizes how much she means to him. As a private revelation, this song contrasts sharply with his earlier blustering speeches about the ungrateful "thing" that he "created out of the squashed cabbage leaves of Covent Garden." It also contrasts with Higgins's deliberate performance of nonchalance a moment later, when Eliza comes back. When she enters the house, catching him as he listens to her voice on the recording machine, he pulls his hat over his face, sinks deeper into the chair, and inquires, "Eliza, where the devil are my slippers?" presumably to show his cool and incorrigibility. The transparent lyrical Higgins of "I've Grown Accustomed to Her Face" is gone (fig. 25).

But it's a tricky business with that hat. Intended to save face, it also hides his face. Perhaps Higgins knows that he has no control over his expression right now and doesn't want to be seen in such a vulnerable state. If that's the case, then Higgins's gesture implies restraint, which makes him almost as transparent with the hat over his face as he was a moment earlier, in the midst of his solitary lyrical musings.

But whether we read Higgins's body language as indicative of restraint (and thus transparency) or nonchalance (and thus performance put on for Eliza), nothing is the same after his song and the glimpse into his soul that it has allowed us. We know how he "really" feels about Eliza, and so does he. The moment of direct access brought about by the song has changed the dynamic of the story. Of course, the story is over; this is the last scene. But, as critical responses to *My Fair Lady* show, we continue to think about what will happen next between Hig-

FIGURE 25. Rex Harrison and Audrey Hepburn in George Cukor's *My Fair Lady* (Warner Bros., 1964).

gins and Eliza, speculating about their relationship in light of what we've just learned.[7]

Theater critics actually have a special term for a song "in which the main character has some kind of revelation or undergoes a major emotional moment that brings the musical to a climax." They call it an eleven-o'clock number—a "holdover from the days when all musicals started at 8:30 PM and had to have a climactic song around 11:00, because it was desirable to have audiences leave not long after 11:00."[8] Higgins's "I've Grown Accustomed to Her Face" is one well-known example of an eleven-o'clock number; "Rose's Turn" from *Gypsy* is another.[9] According to Mark Sheskin (the cognitive psychologist and theater enthusiast who introduced me to this term), some recent musicals have experimented with moving up their eleven-o'clock numbers, thus making the protagonists undergo their epiphanies in earlier acts.[10] But wherever epiphanies are to be found, we can safely say that although not all musicals have eleven-o'clock numbers, all eleven-o'clock numbers are instances of embodied transparency.

So far I've kept it simple. To illustrate the neat three-way division—song-as-performance, song-as-conversation, and song-as-thought—I have used examples from musicals made in the 1950s and 1960s. Things got more complicated, however, in the 1970s and 1980s, particularly in the work of Stephen Sondheim. Sondheim seems to delight in blurring boundaries among the three, introducing embodied transparency into songs-as-conversations, and performance into songs-as-thoughts. (Actually, things weren't that simple in the 1950s and 1960s either. Classic musicals experimented with crossing these boundaries, too. It's just that this experimentation became more common in recent decades.)

Think of the famous "Franklin Shepard Inc." number at the end of the first act of Sondheim's *Merrily We Roll Along* (1981), when Charley Kringas, Frank Shepard's collaborator and friend, has an emotional breakdown during a live television broadcast. An interviewer asks Charley how he and Frank work together (Charley as a scriptwriter, Frank as a composer), but instead of giving a cute or pat answer, Charley bursts into an angry complaint about the disintegration of their collaboration and friendship. It's clear that Charley is as shocked about his outburst as everyone else around him ("Oh, my God, I think it's happened,/Stop me quick before I sink"), but he can't stop venting his pent-up grief and frustration over what Frank's selling out has done to their professional and personal relationship.

"Franklin Shepard Inc." is an eleven-o'clock number moved up to the beginning of the play. Though, as Sheskin notes, "this is an interesting case because the musical is told in reverse order. The number occurs towards the beginning of the audience's night, but at the chronologically correct time in the story."[11] That is, on some level we can still think of this song as an emotional upheaval *concluding* the play.

What makes "Franklin Shepard Inc." unusual as a case of embodied transparency is that it is simultaneously a song-as-thought, a song-as-conversation, and (though to a lesser degree) a song-as-performance. Charley is *experiencing* the anger and grief that he is singing about; he is *talking* to Frank and their interviewer; and he is *performing* (for this is a live TV broadcast, and, at least in the beginning of his song, when Charley is more in control of his feelings, he makes a lovely show of evoking the sounds of a piano and a typewriter).

Here is another example of Sondheim's taking a song-as-thought—and a striking instance of embodied transparency—and complicating it with elements of performance. When in *Sunday in the Park with George* (1984), George Seurat's model and mistress, Dot, gets tired of posing, she steps out of her rigid Victorian dress as one would out of a closet and prances around the park in her underwear, singing about her career as a model and her love for George (fig. 26).

On the surface this is a brilliant visual exploitation of our mind-body dualism in the service of embodied transparency. We get direct access to Dot's mind via her song while other characters in the play (George and the people walking through the park) get none: they are stuck with the body. All they can see is the motionless Dot, encased in her cumbersome dress, looking out at the water, posing for her exacting lover (fig. 27).

FIGURE 26. Dot (Bernadette Peters) next to her dress in Stephen Sondheim's *Sunday in the Park with George* (Broadway, 1984).

FIGURE 27. Dot (Bernadette Peters) poses for George (Mandy Patinkin) in Stephen Sondheim's *Sunday in the Park with George* (Broadway, 1984).

Yet something else is happening here. As Dot is singing about what it means to be a professional model—one gets no "respect," "attention," or "connection" during her life but "some more public and more permanent expression of affection" after death—she keeps striking mock poses for her imaginary audience. These are poses that George would have no use for in his paintings but that *other people* may associate with modeling.

Are these the imaginary spectators who give her "no respect" while she is alive? Perhaps they don't understand George's art either—if they think that he is the kind of artist who needs his models to assume these ridiculous postures. Yet Dot engages these people in her song, giving them what (she thinks) they can immediately appreciate—a sexy girl in frilly underwear—while also knowing that they (or other people who think like them) may reevaluate their present biases in the future (i.e., after both George and Dot are dead and his art gets the recognition it deserves). Dot thus plays both with their current perceptions of artists and their models and their future realization that modeling for the great Seurat must have been a different experience and called for a very special kind of woman.

There is, in other words, a strong element of performance in Dot's transparency. Her song is still clearly a thought—the empty dress standing in for her visible body doesn't let us forget that the disrobed Dot is singing out her "innermost" feelings—but it's also a complex act of mind-reading: giving people what they (presumably) want now in order to shape their future views of her and George. Dot is transparent and manipulative at the same time; Sondheim is brilliant at such mind-reading mashups.

We find a similar dynamic of mixed mind reading throughout Rob Marshall's movie *Chicago* (2002), the revision of John Kander, Fred Ebb, and Bob Fosse's stage musical of 1975. Roxie Hart's number "Funny Honey" is simultaneously an expression of her private thoughts about her husband, Amos, and her imaginary performance in front of a nightclub audience (fig. 28). And when, later in the play, Amos Hart sings "Mister Cellophane," he is both reflecting in private on his social insignificance and performing these sad private thoughts onstage (perhaps the same imaginary stage that his wife inhabits in her dreams).

I believe that Roxie and Amos are still transparent in these scenes because their songs-as-performances are contained within their songs-as-thoughts. Take Roxie. We know exactly what's on her mind. She is thinking that her husband is a sap, and she is imagining how she would look were she to sing about her sap husband in a glamorous nightclub. But even though she hopes her imaginary spectators adore her, she can't be absolutely sure. So while Roxie is letting us in on her most private feelings, she is working hard to put her audience into a certain state of mind: to woo and seduce them. Once more, the musical stage opens up an opportunity for a wonderfully mixed mind-reading experience: social complexity inside transparency.

FIGURE 28. Renée Zellweger as Roxie Hart and John C. Reilly as Amos in Rob Marshall's *Chicago* (Miramax, 2002).

What Is It about Theory of Mind and Singing?

Playing with a convention does not mean doing away with it. If anything, it means just the opposite. Discovering that a song that is supposed to be a thought is also a performance is titillating precisely because we expected transparency. And this expectation is as strong as ever. Our default assumption still is that if a character sings a song that is not clearly marked as a performance (as in a backstage musical) or as a conversation, she is revealing private thoughts and feelings.

But why should it be so? This is a question that scholars of the musical don't ask but that, as a cognitive cultural critic, I must ask. Is there some kind of a special relationship between singing and theory of mind? What is it about a body caught in the act of singing that makes it easy for us to believe that we have direct access to the character's mental state?

If you tell me that this is a mere convention and that we stick with it because it has been around for so long, I have to disagree. I disagree because I am struck by how persistent and cross-cultural the connection between singing and transparency is. In fact, this persistence makes me wonder about the special physical and mental effort that the act of singing requires. Can it be that the awareness of this effort influences the listeners' perception of the singer's emotional transparency? Let us consider this argument in detail.

First, think of other genres in which singing is associated with direct access to thoughts and feelings. So far we have discussed the musical, but, of course, there is also opera. There the default expectation is that characters sing their hearts out for significant periods of time, and, as opera scholars have noted, nobody else onstage can hear them when they do.[12] Moreover, that expectation holds not just for the European operatic tradition. For instance, it is the same in Chinese opera, in which *qu* songs generate moments of intense lyricism by expressing characters' innermost feelings while nobody else onstage can hear them.

Two things are particularly striking about the embodied transparency in the Chinese opera: first, qu songs are full of references to other well-known cultural narratives; second, the body language of actors is highly stylized. In other words, it may seem difficult for characters to appear spontaneous in their emotional expression when they are constantly evoking other songs and poems and going through complex sequences of elaborate symbolic gestures. Yet, as a scholar of classical Chinese literature, Kang-i Sun Chang, observes, out of those verbal and physical references emerges an emotional display that spectators consider particularly delectable.[13] The actors are expected to achieve such a mastery over the stylized body language and literary heritage that they become second nature. The spontaneous emotions of characters thus "shine through" the formulaic gestures and abundant literary and historical quotes.[14]

So when in *The Story of the Western Wing* (the thirteenth-century classic by Wang Shifu), Student Zhang sees beautiful Oriole walking at some distance in a monastery garden and is immediately smitten by her, his song certainly represents a spontaneous outpouring of his private feelings, even though it also happens to demonstrate his knowledge of history, mythology, and drama. Zhang is dazzled and "speechless," yet he still speaks learnedly of a mythical realm of bliss, the "Tushita Palace," and "the heaven of Separation's Regret," an imaginary abode of thwarted lovers, typically evoked by drama of the period:

Stunning knockouts—I've seen a million;
But a lovely face like this is rarely seen!
It dazzles a man's eyes, stuns him speechless,
And makes his soul fly away into the heavens.
She there, without a thought of teasing, fragrant shoulders bare,
Simply twirls the flower, smiling.
This is Tushita Palace,
Don't guess it to be the heaven of Separation's Regret.
Ah, who would have ever thought that I would meet a divine sylph?
I see her spring-breeze face, fit for anger, fit for joy,
Just suited to those flowered pins pasted with kingfisher feathers.[15]

Neither Oriole nor anybody else in the garden can hear Zhang when he sings about Oriole's "spring-breeze face" and compares the monastery to the Tushita Palace. Were they to hear him, they would be offended by his forwardness. This contrast between ordinary speech and the song as a private thought is emphasized a moment later when Zhang stops singing, turns to a monk who happens to be nearby, and blurts ecstatically, "Monk! Guanyin just materialized!" to which the monk replies reasonably, "You're babbling. This is the young daughter of Chancellor Cui" (121).

Wang also includes songs meant as performances (when Zhang and Oriole exchange improvised couplets [140]) and songs meant as conversations (when Oriole's servant, Crimson, advises Zhang not to fear Oriole's mother [130]). In other words, *The Story of the Western Wing* employs the same three strategies for introducing songs into a play as do classic Western operas and musicals. Qu songs are clearly predominant, however. Again, in this respect, Chinese opera prefigures Western opera, in which the majority of songs render characters transparent.

Consider now another representational tradition, roughly coeval with the period to which *The Story of the Western Wing* belongs: medieval European romances, with their frequent insertions of song into narrative. Once more, some of these songs are public performances—as when a minstrel entertains courtiers at a fashionable gathering—but the majority are used to express private romantic longings.

In 1993 the literary critic Maureen Barry McCann Boulton examined "every lyric insertion" in each of the seventy-two French narratives composed between 1200 and 1400 that contained such insertions. She found that the "attribution of a song to a character for the purpose of expressing or analyzing his (or occasionally her) sentiments . . . eventually became the most successful application of the device." Particularly when used as part of the interior monologue, the songs were "the principal device for revealing the inner thoughts of a character."[16]

Boulton did not write her book merely to confirm the view that songs in French medieval stories expressed the hidden emotions of characters. In fact, it was the opposite. She wanted to question this view by reminding her readers that many of these songs contained extended references to other well-known performances. As she saw it, what this tradition of recycling and conscious referencing must mean is that we cannot continue to regard those lyrical insertions as spontaneous outpourings of characters' private thoughts.[17]

You can see why I can't quite agree with Boulton's argument. Think again of the Chinese opera. Its qu songs are full of references to other songs, myths, and historical narratives, yet this does not render them less lyrical. That is, to use the definition of lyricism offered by Chang, qu songs still contain a "sustained expression of [characters'] emotions, felt in the present."[18] So, to return to Boulton's analysis of medieval lyrical insertions, it seems to me that the abundance of references and direct borrowings does not by itself invalidate our association of a song with direct access to a character's mind.

This brings us to our everyday freestanding song. Our default expectation about such a song, an unspoken rule, is that it reveals the feelings of the persona assumed by the singer. (The word *persona* is crucial here: I do not claim that a song gives us access to the mind of the singer—only to the character assumed by her for this particular performance.)

There is, of course, an important exception to this rule: a narrative ballad. Instead of focusing on mental states of the protagonist ("Oh, I believe in yesterday"), it tells us about the relationship between several characters, whose actions we can follow but whose minds we can't read, or, at least, not right away. Thus Bob Dylan's "The Lonesome Death of Hattie Carroll":

William Zanzinger killed poor Hattie Carroll
With a cane that he twirled around his diamond ring finger

At a Baltimore hotel society gath'rin'
And the cops were called in and his weapon took from him
As they rode him in custody down to the station
And booked William Zanzinger for first-degree murder . . .

Because the singer explicitly sets out to *tell us a story*, a narrative ballad is a stand-alone equivalent of Higgins's "Why Can't the English?" That is, it's a song as a speech—not a song as a thought—even though it may contain the singer's mental states, as well as the mental states he attributes to his listeners, as in the last three lines of each stanza of "The Lonesome Death":

But you who philosophize disgrace and criticize all fears
Take the rag away from your face
Now ain't the time for your tears.

So, if we imagine a full continuum of embodied transparency induced by singing, it will look as follows. On one end there will be lyrical songs implying complete transparency of the character assumed by the singer, such as "I'll Never Smile Again." On the opposite end there will be songs with the least transparency: narrative ballads featuring several nontransparent minds, as in "Raggle Taggle Gypsy-o."[19] In between there will be operatic arias, songs from musicals, lyrical insertions from medieval romances, and an endless variety of other musical numbers. Within each of these categories some songs will gravitate toward one end of the spectrum and others toward the opposite end. *As a whole*, however, the spectrum will be weighted heavily toward transparency. That is, in spite of the range of differences within the categories, each category as a whole will have more songs creating contexts for embodied transparency than those featuring unreadable minds.

Why should it be this way? I am returning now to my earlier suggestion that there is something about the act of singing that makes it easy for us to believe that the song is revealing the true feelings of the persona assumed by the singer. At this point I can only speculate why it might be so.

The starting point for my speculation is the testimony of performers whose roles demand that they act and sing simultaneously. As the actress Evan Rachel Wood, who started out in musical theater, puts it, "The hardest part is singing live and acting at the same time. You have to sing emotion and sing the song well."[20] Opera singer Natalie Dessay concurs: "It's almost impossible to sing and really act at the same time."[21] The fact that performers do frequently combine singing and acting may obscure the hard work and compositional skills involved. As Sheskin points out, the best musical writers, such as Sondheim, know how to construct a scene to make the sudden eruption into song easier for an actor, that is, how to gradually ratchet up emotions to such a pitch that at some point switching to a different vocal mode feels almost natural (or even necessary).[22]

One possible explanation for why simultaneous singing and acting might be difficult is that singing is effortful. It requires a sustained shift in vocal tone, pitch, and rhythm. So perhaps we intuitively expect greater access to the "true" feelings of people when they are in this changed vocal mode because we are aware that they are already expending a lot of energy and thus may not be able to shoulder the additional cognitive load of pretending. This is, obviously, a highly speculative suggestion, but, even if it is wrong, the strong association between singing and transparency is still an intriguing cognitive phenomenon that we need to address.

And meanwhile narrative ballads, songs as performances, and songs as conversations intuitively rely on this association and subvert it. Take Lou Reed's "Walk on the Wild Side" or Billy Joel's "Piano Man." Were I to put in words the pretend-game that the personas assumed by performers who sing such songs play with our mind-reading adaptations, it would go like this: "Yes, I am in the vocal mode that must give you immediate direct access to my thoughts and emotions. Yet something else seems to be going on. Continue looking around. In spite of the evidence of your senses, you can't really know what I am up to. You will have to figure out what I am thinking and feeling by further engaging with the social situation at hand, that is, by paying attention to other minds around me and to my perception of those other minds."

So without concluding that singing is "all and only" about mind reading, I still say that, at least to some significant extent, the pleasures of singing and listening to songs are sociocognitive pleasures. To the degree to which mind reading is a crucial cognitive feature of our social species, different musical numbers offer us different ways to read minds.

We will never get enough of them—neither minds nor songs.

NINE

||||||||||||||||||||||||||||||||||||||||||||||||||||||||||||||||||||||||||||||||||||||||||||||||||||||||||||||||||||||||||||||||||||||||||||||||||||||||||||||||||||||||

In which bubbles are blown and card castles are built / rules for hating abstract art are delineated / extreme measures are taken, and emotions run high / benevolent sons-in-law step into the breach, but some people still sneer / da Vinci lurks / Rubens takes a stroll in the garden / and the author waxes sentimental about a football memoir.

Painting Feelings

Absorption as Transparency

It's warm outside. Spring blossoms brush against the house. Leaning over the windowsill, propping his right hand with his left, a young man is blowing bubbles. Just now a particularly large bubble is trembling at the tip of his blowpipe.[1] The man is holding his breath. The world is standing still (fig. 29).

Soap Bubbles is one of Jean-Baptiste-Siméon Chardin's "paintings of games and amusements" done in the 1730s.[2] His subjects build card castles, sketch, and play knucklebones. They are so completely absorbed in what they do that they are unaware of being watched, and they draw us in precisely with their peculiar obliviousness to our presence, their utter lack of performance.

Absorptive paintings are both irresistible and difficult to create. This is the argument of Michael Fried in *Absorption and Theatricality: Painting and Beholder in the Age of Diderot*. Following the development of French genre painting from the 1730s to the early 1780s, Fried shows

FIGURE 29. Jean-Baptiste-Siméon Chardin, *Soap Bubbles*, c. 1733–34. Oil on canvas, 24 × 24⅞ in.

how artists tried to minimize the self-awareness of art. To resist the "primordial convention that [art is] made to be beheld," they depicted people not aware of the presence of the beholder.[3] He also shows how quickly the established methods of representing absorption would become stale and how desperately artists would cast about for new ways to convince audiences that the people in paintings did not care about their gaze.

Published in 1980, *Absorption and Theatricality* continues to influence art historians and cultural critics. I personally find its argument exciting because I see absorption as embodied transparency (with one small exception, to be discussed in the next chapter).[4] We know exactly what the young man in Chardin's painting is feeling. His whole attention—his whole mind—is on that bubble. And we also know that as soon as the bubble bursts (in just a matter of seconds) his mind will move to other things, and he won't be transparent anymore. Remember the rule of transience?

I am running ahead of myself. Before we get to see how the rule of transience and the rule of contrasts play out in absorptive paintings, I want to consider, if only briefly, the larger question of the relationship between theory of mind and art. And, even before that, I want to say that another reason I am excited about *Absorption and Theatricality* is that Fried had neither background nor interest in cognitive science when he wrote his book. This is important to me because I am always happy when I find studies in art history, literature, or cultural studies compatible with research in the cognitive sciences. If we are all in the business of figuring out how the mind works, then arriving at complementary conclusions while starting off from very different disciplinary perspectives is a good indication that we really are on to something.

Theory of Mind and Art: Whose Mind Do We Read When We Look at a Painting?

It turns out that even deciding whether an object can be considered art is a mind-reading decision because it depends on our understanding of the *intentions* of the person behind the object.

Cognitive psychologists Susan Gelman and Paul Bloom have found that, already at the age of three, children pronounce a blob of paint to be a painting only if there was a particular mental state behind its creation: "We showed the children a blob of paint on canvas . . . and either said that it was created by a child who accidentally spilled his paint or by a child who used his paint very carefully. As predicted, this made a difference: When told that it was an accidental creation, children tended to later describe it using words like 'paint'; but when told that it was created on purpose, children tended to describe it as art—as 'a painting.'"[5]

As Bloom observes, the traditional view in developmental psychology was that children name pictures they draw based on their appearance. That is, "for a child, the word 'airplane' should refer to something that looks like an airplane." However, as Bloom and his graduate students have established in a series of studies, the

> names aren't driven by what pictures look like; they are chosen on the basis of the pictures' histories. . . . Even three-year-olds would name their pictures based on what they were intending when they created it. We also found that the same holds for pictures that other people draw. If a three-year-old watches someone stare at a fork and draw a scribble, she will later name the scribble "a fork"; if the same scribble was created while the person looked at a spoon, she will call it "a spoon." In more recent studies [we] found that even 24-month-olds are sensitive to a drawing's history when deciding what to call it.[6]

Different eras in art have different ways of experimenting with our intuition about the importance of intention behind an object. Marcel Duchamp pushed this intuition to what must have felt like its logical end point in 1917 with his *Fountain*, a urinal submitted to the exhibition of the Society of Independent Artists. I am writing this having just come back from a retrospective of Francis Alys at the Museum of Modern Art. Alys's filmed endeavors include "pushing a block of ice around Mexico city until it melted; walking around the same town with a large pistol to see what would happen (he was arrested after eleven excruciatingly long minutes); [and] enlisting five hundred people, in Peru, with as many shovels to move part of a mountainous sand dune."[7] Because the relocation of the sand dune was intended as art and not, say, a public-works project, it can be considered an object of art. Not that it necessarily will be—many other culture-specific stars have to align in certain lucky ways—but a starting point (i.e., the intention) is in place.

I like Alys's work for the same reason I like the work of many other conceptual artists: it makes me see the world differently. I wouldn't characterize it as beautiful, though. This distinction is important because, when I say that attributing a certain mental state to a person behind an object is a prerequisite for considering it a work of art, I am not talking about aesthetics. There is a gap between calling a blob of paint a "painting" because it was created intentionally and figuring out its aesthetic value, and I am not bridging this gap here.[8] My focus is on the sociocognitive dynamics of art. I want to understand whose minds we read and why, when we encounter works of art, particularly—given that ultimately I will come back to embodied transparency—those depicting human bodies.

Consider an art museum as one particular haunt of greedy mind readers. Its existence depends on complex acts of mind attribution involving the curators, the public, and the artist. To touch on just one aspect of this mind attribution, the curators recognize the intention of the artist to create an art object and hope to elicit certain mental states in people who will come across it in the museum. In some cases the curators insist that an object not initially intended as a work of art be perceived as such by the public (e.g., home appliances that at a later period, or in a different context, are seen as peculiarly stylistically expressive).

Moreover, once an object—say a painting with human figures in it—is exhibited in a museum, its reception will necessarily involve additional attribution of mental states. The main question at that point is who gets the lion's share of mental states—the people in the painting (i.e., its subjects), the artist, or the audience—and the answer depends both on the content of the artwork and on the context in which it's viewed. Let's look at these three options—the subjects, the artist, and the audience—separately.

By portraying bodies, artists portray minds. That is, we make sense of a painting by attributing mental states to the people depicted in it. We do so by observing their body language, by interpreting the body language of other people in the painting who interact with them, and by turning to cultural narratives that exist independently of the painting but may have influenced it. Although we do some of it consciously—we remember, for example, the biblical story behind da Vinci's *The Last Supper*—much of it happens outside of our conscious awareness.[9]

If a painting actively prevents us from attributing mental states to its subjects, we may turn to the artist. Think, for instance, of surrealist paintings that depict human figures, or claim to do so in their titles, but that make it impossible to read the body language of their subjects, such as de Chirico's *The Uncertainty of the Poet*; Ernst's *Family Excursions, Approaching Puberty, Woman, Old Man, and Flower, Loplop Introduces a Young Girl,* and *The Robing of the Bride;* or Picasso's *Young Tormented Girl* and *Silhouette of Picasso and Young Girl Crying.*[10] Quite often in such cases, we try to learn as much as we can about the intentions of the artist and about the life events and aesthetic influences that might have shaped these intentions. What did he or she mean, we ask, focusing most of our mind-reading energies on the artist.

The third possibility: we can speak about a mood or feeling that a painting creates in *us,* which means that we try to make sense of it by attributing mental states to ourselves. This works even in those situations, and perhaps particularly in those situations, when a painting is baffling to the point of being irritating. Let's say I saw an abstract piece that I absolutely hated: I didn't understand it, I couldn't see anything in it, I felt like I wasted my time, and I decided that I don't ever want to look at anything by this artist again. My negative response seems to indicate a lack of engagement with the painting, but note all the references to mental states in my diatribe: all this talk about hating, understanding, feeling, wanting, and deciding. Where is it coming from, and what is it doing here?

These references to thinking and feeling must mean that I am still making sense of the painting by exercising my theory of mind—except that, unable to read the minds of the subjects of the painting and unwilling to read the mind of the artist, I have transferred my mind-reading impulse onto myself and focused on the wishes, attitudes, and intentions that the painting prompted in *me.* It is as if we approach each painting ready and eager to attribute states of mind, and if something prevents us from attributing them to the subjects of the painting, we turn with the same eagerness to the artist and start thinking about her mind, and if we can't do that, we begin to attribute mental states to ourselves. *We will read minds,* our behavior on such occasions seems to say; *we know that by approaching a work of art we enter into an environment in which there are minds to be read, and we will do so, come what may!*[11]

I first came across Lyonel Feininger's *The Green Bridge II* in a Whitney Museum exhibit, on a rainy day. It took my breath away. I still don't know why I liked it so much—it might have something to do with its colors and view of the world—but all I wanted to do was to keep looking at it or, even better, keep coming across it unexpectedly after looking at other things. (That's why it's now the frontispiece of this book.) It is as if I wanted to experience again and again that feeling of having the wind knocked out of me. My gut reaction to something in the painting expressed itself as a keen awareness of my mental state and a wish to keep myself in that particular mental state.

Yasmina Reza's play *Art* depicts mind reading prompted by a painting that doesn't allow one to attribute mental states to anybody in the painting or to the artist: a white canvas with three off-white lines in the middle. So the three protagonists of *Art* spend the whole play attributing complex mental states to themselves and each other. The painting becomes a catalyst for their anxieties about social status and friendship. In the following excerpts, typical for *Art* in this respect, note with what gusto Marc, Serge, and Yvan discuss each other's mental states while ostensibly talking about the painting.

> MARC. You paid two hundred thousand francs for this shit?
>
> .   .   .   .
>
> SERGE. You have no interest whatsoever in contemporary painting, you never have had. This is the field about which you know absolutely nothing, so how can you assert that any given object, which conforms to laws you don't understand, is shit?
>
> MARC. Because it is. It's shit. I'm sorry.
>
> *Serge, alone.*
>
> SERGE. He doesn't like the painting.
>
> Fine . . .
>
> But there was no warmth in the way he reacted.
>
> No attempt.
>
> No warmth when he dismissed it out of hand.
>
> Just that vile, pretentious laugh.
>
> A real know-all laugh.
>
> I hated that laugh.[12]

And on a different occasion:

> YVAN. We had a laugh. . . .
>
> MARC. He wasn't laughing because his painting is ridiculous, you and he weren't laughing for the same reasons, you were laughing at the painting and he was laughing to ingratiate himself, to put himself on your wavelength, to show you that on top of being an aesthete who can spend more on a painting that you earn in a year, he's still the same old subversive mate who likes a good laugh.[13]

Indeed it seems that when it comes to abstract art, viewers attribute mental states mostly to the artist and/or to themselves.[14] This is in contrast, say, to absorptive paintings, which seem to confine our mind reading to the mental states of their subjects.

That, too, may change, though, depending on the context in which one views them. For the sake of this discussion I have considered separately the three types of mind-reading attribution that engagement with a work of art may prompt in a viewer. But in reality it's often a balance of the three, while a specific historical, academic, or personal context defines what kind of mind attribution takes precedence. For instance, I am sure you can think of different contexts in which you would attribute mental states mostly to the subject of *Mona Lisa,* mostly to da Vinci, or mostly to yourself.

Moreover, artists may intuitively experiment with our theory of mind by working with a genre that is expected to stimulate a particular kind of mind reading and subverting this expectation. Think, for instance, of seventeenth- and eighteenth-century conversation pieces such as Hogarth's *The Assembly at Wanstead House,* Zoffani's *Charles Towneley in his Sculpture Gallery,* or Devis's *Sir George and Lady Strickland in the Park of Boynton Hall*—paintings that depict well-to-do families and, sometimes, their friends engaged in pleasant social activities. Conversation pieces typically contain enough rich information about the mental states of their subjects for us to dwell on those mental states contentedly and at length without wondering about the artists' intentions.

Now think of Rubens's *Self-Portrait with Hélène Fourment in Their Garden* (1630–31?) and Velazquez's *Las Meninas* (1656), also conversation pieces. Both artists manage to subvert this mind-reading pattern by integrating themselves into their paintings. Rubens leans forward from behind Hélène, looking directly at the viewer (fig. 30); Velazquez pauses next to the easel with the brush in hand, his eyes resting on the viewer (or, depending on which interpretation of *Las Meninas* you prefer, on the king and queen of Spain).[15]

In each case we almost have no choice but to start thinking about the intentions of the artist even though the genre of the painting—conversation piece—typically doesn't call for attributing mental states to anybody outside the painting. I suspect that our mind-reading expectations are similarly short-circuited (while our interpretations enriched!) when we come across unexpected integrations of artists into their paintings in other genres. (Of course, this applies not just to paintings. How does it

FIGURE 30. Peter Paul Rubens, *Self-Portrait with Hélène Fourment in Their Garden,* c. 1630. Oil on panel, 51.6 × 39 in.

make you feel when you recognize that a portly man crossing the street in an enthralling mystery movie is Hitchcock himself?)

But no matter what genre convention a given work of art subverts, or in what context we happen to view it, there is no engagement with a work of art that is somehow theory of mind–free. From recognizing something special about the intent of the person drawing animals in charcoal on the wall of a cave thirty-two thousand years ago,[16] to walking in puzzlement around an object cordoned off by a wire on the floor of an art museum, to following the gaze of a young woman who is observing an interaction between two older women in a conversation piece, art has always been and will remain a product of a culture of greedy mind readers.

This man wants his bubble to last another moment. That woman is lost in her studies. That boy hopes to add another card to his card castle without having it tumble down. This woman is startled by the arrival of her son. This man is asleep.

Absorptive paintings focus our attention on the mental states of their subjects *and* provide us with direct access to their minds. (The two don't have to go together. Think again of *Mona Lisa:* the more we focus on her mental state the less we know about it.) Yet even with absorptive paintings and their lure of embodied transparency, our attention does not stay confined to the mental states of their subjects for long. Art critics and art historians ensure that it doesn't.

By convincing us to think of absorptive paintings in terms of their subjects' attitude toward the beholder, Fried shifted our mind-reading balance. To the extent to which we perceive the sleeping hermit (fig. 31) as not noticing the beholder (us), we read our own mental states as part of our engagement with the picture. That is, instead of merely seeing the hermit as asleep, we are now also conscious of our awareness of not being noticed.

This is what art criticism does—it introduces more mental states into our perception of an artwork.[17] In the case of absorptive paintings the very act of recognizing a particular piece as "absorptive" immediately redistributes our mind-reading energies three ways: we think of the mental state of its subject (the hermit is asleep), of our own mental state (we are aware of not being noticed), and of the mental state of the artist (Vien must have intended to depict the hermit as completely oblivious to the outside world).

At certain historical junctures this distribution of mind-reading effort may become a zero-sum game involving the subject of the painting and the artist. For instance, if we start finding the subject's absolute focus on what he is doing less convincing—for instance, if we are not sure we can read his mind—we focus more and more of our mind-reading attention on the artist. We begin to speculate not just about the artist's intention to render her subject absorbed but also about her intention to anticipate her audience's skeptical response to her work.

FIGURE 31. Joseph-Marie Vien, *The Hermit,* 1750. Oil on canvas, 87 × 58 in.

I discuss this zero-sum game below, when I turn to the inevitable moment when a culture becomes aware of the seemingly reliable context for embodied transparency and begins to treat such a context with skepticism. But first let us see how the rules of transience, contrast, and restraint fare with absorptive paintings.

Absorptive paintings cultivate transience. Here is a long quote from Fried, in which he comments on the artist's "uncanny power" to make us think that the moment of absorption will not last, that what we are seeing right now is about to change:

> Chardin's paintings of games and amusements, in fact, all his genre
> paintings, are also remarkable for their uncanny power to suggest the
> actual duration of the absorptive states and activities they represent.
> Some such power necessarily characterizes all persuasive depictions of
> absorption, none of which would *be* persuasive if they did not at least
> convey the idea that the state or activity in question was sustained for
> a certain length of time. But Chardin's genre paintings, like Vermeer's
> before him, go much further than that. By a technical feat that almost
> defies analysis—though one writer has remarked helpfully on Char-
> din's characteristic choice of "natural pause in the action which, we
> feel, will recommence a moment later"—they come close to translat-
> ing literal duration, the actual passage of time as one stands before
> the canvas, into a purely pictorial effect: as if the very stability and
> unchangingness of the painted image are perceived by the beholder
> not as material properties that could not be otherwise but as manifes-
> tations of an absorptive state—the image of absorption in itself, so to
> speak—that only happens to subsist. The result, paradoxically, is that
> stability and unchangingness are endowed to an astonishing degree
> with the power to conjure an illusion of imminent or gradual or even
> fairly abrupt change.[18]

Chardin's paintings intuitively play up our mind-reading uncertainties. They make us believe that we have direct access to these people's minds *now* by making us expect to lose this access any second. They effec-tively reinforce our anxious suspicion that other people's minds are never transparent by presenting *this* moment of transparency as an exception, an accident, a fluke. By doing so, they make us value this fluke—they encourage us to seize the moment and to look and look and look at it while it lasts.[19]

Note something else about the transience of absorptive moments. Whereas in fiction we can talk separately about the rule of contrasts and the rule of transience, in absorptive paintings the two blend together. Visual transience implies contrast. If a painting succeeds in creating the impression that in a couple of seconds the absorbed person will not be absorbed anymore—so we won't know what she is thinking—it means that on some level we are already imagining that person as not transparent. Her present transparent self is contrasted with her future nontransparent self.

While the rules of transience and contrast are central to absorptive paintings, the rule of restraint is not. In fact, it seems to be largely absent in stand-alone paintings, as opposed, that is, to sequential art, such as comic strips, graphic novels, or other forms of visual art in which a narrative unfolds in a series of images. When I try to think of restraint as embodied transparency in stand-alone visual representations, what mainly comes to mind are book illustrations accompanying the moments of embodied transparency in fictional narratives. (And, of course, these are not exactly stand-alone pictures either, since they are embedded in books. Let's agree, then, for the sake of this argument, that stand-alone means "distinct from a series.")

Imagine, for example, an illustration facing the page from *Pride and Prejudice* in which Mr. Darcy is described as trying to conceal his anger and disappointment at Elizabeth's rejection: "His complexion became pale with anger, and the disturbance of his mind was visible in every feature. He was struggling for the appearance of composure, and would not open his lips, till he believed himself to have attained it" (129). Such an illustration would not be able to capture restraint on its own. Instead it would show Mr. Darcy as looking angry or just focusing intently on Elizabeth's face (see fig. 3 as an example and imagine that this is a stand-alone picture, not a scene from a movie).

We would *read* restraint into the picture because we have the text in front of us, but were we to show that picture to someone not familiar with *Pride and Prejudice* (or even someone who is familiar with it but doesn't remember every detail of that specific moment), that person would not see Mr. Darcy as struggling with his emotions. Because restraint is dynamic yet brief—unfolding in time yet over almost instantaneously—portraying a character who tries to look composed while his mind is disturbed presents an artist with a serious challenge.

"Deliberate and Extraordinary Measures"

We already saw this happen with another eighteenth-century artifact: the novel. When a cultural setting emerges as an established context for embodied transparency, that setting loses credibility. Occasions for the unpremeditated display of feelings become occasions for more devious performance. Once the strategy of bringing a character to the theater was recognized as a convention for letting other characters read his mind, theater became a place where Robert Lovelace (from Richardson's novel *Clarissa*) could fake his deep emotional engagement with a play to impress whoever would care to observe him.

It happened with absorptive paintings, too. Chardin's canvases of the 1730s, as well as Chardin's, Greuze's, Van Loo's, and Vien's works of the 1750s, depicted people so caught up in praying, playing, sketching, learning difficult lessons, blowing bubbles, grieving, rejoicing, listening raptly to charismatic speakers, or simply sleeping, as not to be aware of being watched. By "negating the beholder's presence" these paintings riveted their audiences.[20] The sight of people so absorbed in what they are doing that they are unable to put on any special body postures or facial expressions was mesmerizing. By the early 1760s, however, the subject matter had worn itself thin. It became increasingly difficult for artists to use the established contexts of absorption (reading, praying, sleeping, etc.) to convincingly exclude the beholder from the picture.

The contexts for embodied transparency were becoming conventional and as such—just like theater in fiction—vulnerable to subversion.[21] Imagine a portrait of a woman not merely blowing bubbles but Blowing Bubbles—intentionally engaging in an activity that is *supposed* to be absorptive—*performing* unselfconsciousness for the beholder and thus completely defeating the original purpose of the endeavor. This is a hypothetical image, but I suspect that the artists had in mind something along these lines when they felt that they could no longer rely on the established contexts of absorption.

So, as Fried tells us, "deliberate and extraordinary measures came to be required in order to persuade contemporary audiences of the absorption of a figure or group of figures in the world of the painting."[22] One such measure involved ratcheting up the drama. Greuze's *Le Fils ingrat* (1777) and *Le Fils puni* (1778) depict a family so distraught over the rebellion of an ungrateful son and the resulting early death of his father that it is obvious that none of them would be able to gather their wits enough to look about themselves and realize that they are being observed.

Another measure involved opening a painting "to a number of points of view other than that of the beholder standing before the canvas."[23] In David's *Belisaire* (1785) the "off-center perspective [places] the beholder to one side of the painting, away from [the central] figure of [Belisaire]."[24] (Compare this to *Soap Bubbles*, in which the beholder is squarely facing the subject of the painting.) *We don't care what you see, how you see it, or whether you can see it at all,* such a perspective seems to say to the viewer; *you may just as well not be there.*

Yet another way to create the illusion of absorption was to make the titular character blind, as in Vincent's *Belisaire* (1777), David's *Belisaire* (1781, 1785), Peyron's *Belisaire* (1779), and David's *Homere endormi* and *Homere recitant* (both 1794). The blind protagonist is by default unaware of the beholder.

Fried calls these measures "extreme." Indeed, the need to tap new contexts for absorption sometimes led to strained psychological dynamics within the painting. Consider Vincent's *Belisaire*, in which the protagonist, the famous general, now impoverished and blind, is receiving charity, while a younger officer is looking at him (fig. 32). According to Fried there is something forced about the posture of the young man. He "gazes anxiously, almost mistrustfully, at the sightless eyes of the great general."[25] Does he think that Belisaire is faking blindness? Why should he doubt the old warrior instead of, say, just pitying him?

Fried interprets the young officer's intense gaze as an indication that the artist is desperately trying to make both men seem completely absorbed in the present moment and thus oblivious to the presence of the beholder. The general cannot perform for the beholder because he is blind, and the officer cannot perform for the beholder because he is too preoccupied with figuring out what the blind man is up to. But it is precisely because the officer's attitude is not entirely psychologically con-

FIGURE 32. François-André Vincent's *Belisaire,* 1776. Oil on canvas, 29 × 23 in.

vincing that we infer that it must serve other representational needs (i.e., the need for absorption).

The intensification of drama, the experimentation with different perspectives, and the introduction of blind historical and mythical figures all seem to testify that by the 1770s "the everyday as such was in an important sense lost to pictorial representation."[26] The absorptive charm of such mundane activities as listening, watching, daydreaming was broken. In fact, Fried argues that if we follow "the evolution of David's art between 1780 and 1814," we can trace in it "a drastic loss of conviction in [both] action and expression as resources for ambitious painting, that is, in the very possibility that either could be represented other than as theatrical."[27] In other words, "the persuasive representation of absorption" was something that artists still wanted to achieve, but, at least within the context of that specific period in French art history, their means for doing so seemed to have been exhausted.[28]

What David and other artists perceived as the specific representational crisis is actually an expression of a broader cognitive challenge involved in constructing contexts of embodied transparency. At any given historical moment there seems to be a very limited window of opportunity within which audiences will buy the idea of the complete unselfconsciousness of people engaged in a certain activity or behaving in a certain way. Then the double perspective of the body reasserts itself with a vengeance. Spontaneity begins to feel calculated, sincerity fake, and sentiment sentimental.

But Who Decides Whether Something Is Sentimental?

Let's revisit one extreme measure taken by the French artists in search of new contexts of absorption: the ratcheting up of drama. We really should call it the increase in social complexity. Blowing bubbles, playing knucklebones, and sleeping make for visually compelling images, but there is very little social interaction to any of them. In contrast, making absorption emerge out of such situations as the rebellion and return of an ungrateful son (Greuze's *Le Fils ingrat* and *Le Fils puni*) implies a manifold increase in social complexity.[29] To make the moments of transparency seem brief and spontaneous, artists now have to engage in intricate backstage plotting. They enter the territory of fiction writers (for remember: writers create involved social situations to bring characters to a point at which their bodies reveal their minds), and not everybody is happy with it.

Consider the contemporary responses to Greuze's *La Piété filiale* (1761) (fig. 33). This painting features a paralyzed old man surrounded by his family at the precise moment when they all react emotionally to his interaction with his benevolent son-in-law. Fried recounts Denis Diderot's description of the painting: "The moment . . . chosen by the artist is special. By chance it happened that, on that particular day, it was his son-in-law who brought the old man some food, and the latter, moved, showed his gratitude in such an animated and earnest way that it interrupted the occupations and attracted the attention of the whole family." Diderot "seems almost to be saying that Greuze was compelled first to paralyze the old man and then to orchestrate an entire sequence of ostensibly chance events in order to arrive in the end at the sort of emotionally charged, highly moralized, and dramatically unified situation that alone was capable of embodying with sufficient perspicuousness the absorptive states of suspension of activity and fixing of attention that painter and critic alike regarded as paramount."[30]

If Diderot merely drew his readers' attention to the amusing clash between the spontaneity of the moment depicted in the painting and the intricate planning necessary for achieving this illusion of spontaneity, other contemporaries actually accused Greuze of a "mania for plotting."[31] And later critics went further and taxed him with sentimentality. According to Fried, "for a long time now it has been traditional, almost obligatory, to remark that we, the modern public, no longer find it in ourselves to be moved by the sentimentality, emotionalism, and moralism of much of Greuze's production."[32]

I find this charge of sentimentality extremely interesting from the point of view of theory of mind. It seems that our first response when we initially encounter something that we may later consider sentimental is the excitement about prospective mind reading. We think that we are in a situation in which we have privileged access to certain people's emotions. For instance, we read of a young woman's blushing as she opens a letter, and we assume that we have caught her at a rare, and thus valuable, moment when her body betrays her feelings.[33] But then something happens that dampens this initial excitement.

We find out that this particular instance of privileged access is actually part of a cultural convention. In the case of the blushing young woman, we may learn (for instance, from a professor in class) that we are reading an eighteenth-century *sentimental* novel, in which it is expected that characters blush, pant, pale, and cry. This refocuses our attention on the author instead of the characters. Ah!—we say, as it were—this young woman blushes not because she is painfully self-conscious about the ambiguous social situation in which she finds herself. She blushes because the author, writing *that* kind of novel, feels obligated to make characters blush with some regularity.

So it almost doesn't matter anymore that within the universe of the novel the character may still be transparent. Our perception of privileged access is now devalued. Instead of feeling like brilliant social players

FIGURE 33. Jean Baptiste Greuze, *La Piété filiale*, 1761. Oil on canvas, 21.5 × 25.5 in.

(which we feel when embodied transparency is done convincingly), we feel like patsies manipulated by the author.

Calling something sentimental may thus be a way for us to assert that we do not buy into its promise of privileged mind-reading, that we recognize it as contrived, and that we will look for "truer" feelings elsewhere.

The history of the word *sentimental* itself seems to reflect the same conceptual move from enthusiasm about direct mind-access to disillusionment. The term underwent a change between 1740 and 1820. Originally neutral, "characterized by sentiment," or positive, "characterized by or exhibiting refined and elevated feeling," it acquired a pejorative meaning of "addicted to indulgence in superficial emotion."[34] It is as if when a culture goes through a phase in which it starts distrusting a particular set of ready-made displays of emotion (say, those used by novel-ists), it needs a special term for its distrust.

You can see how within this frame of thinking, it is easy to charge Greuze with sentimentalism. He seems to tell us a real story, but it's a mere pretense because all he really wants to do is display feelings and make them convincing. Everything and everybody in the painting (and it's a well-populated painting!) are but means to this end. Once you start looking at *La Piété filiale* from this perspective, all you see is a "mania for plotting" and "superficial emotion." Greuze now seems to be truly on par with fiction writers—particularly authors of eighteenth-century sentimental novels.

Fried does not settle for the easy tag of sentimentalism in his discus-sion of Greuze. Indeed, the way he approaches these paintings makes his a protocognitive argument, ahead of his time. I have suggested that calling a painting or a novel sentimental is our shorthand way of making a rather complex judgment about its mind-reading dynamics, but Fried made a very similar point as early as 1980, observing that when we talk of eighteenth-century "sentimentalism, emotionalism, and moralism," we don't really explain as much as we think we do. For Fried sentimental-ism is not an end in itself; instead, it fulfills the artist's need to represent absorption: "[We] take those qualities at face value, as if they and noth-ing more were at stake in his pictures; and that we therefore fail to grasp what his sentimentalism, emotionalism, and moralism, as well as his al-leged mania for plotting, are in the service of, pictorially speaking—viz., a more urgent and extreme evocation of absorption than can be found in the work of Chardin, Van Loo, Vien, or any other French painter of that time."[35]

I call Fried's argument protocognitive because I consider absorp-tion to be embodied transparency. When he says that the French artists used techniques that we now call sentimental to plunge their subjects into absorption, I agree but add that, on a larger scale, what they were really after (without knowing it) was embodied transparency. That is, eighteenth-century writers and artists faced the same challenge that writ-ers and artists always face: they wanted to construct convincing repre-sentational contexts for making the body reveal the mind. The fact that we now group some of their methods under the unflattering rubric "sen-timentalism" shows again how quickly those methods become outmoded and how ready we are to suspect that there is an element of performance in any show of sincerity.

But perhaps the negative connotations of the late eighteenth-century term *sentimentalism* show something else, too. Think of how many novels, movies, and songs produced within any recent decade can be easily characterized as sentimental, not in the pejorative sense of the word but in the earlier eighteenth-century sense: as "characterized by sentiment." Not to search too far, even Hornby's football memoir *Fever Pitch* fits the bill with its emphasis on bodies caught in spontaneous emotional reaction to the game. That kind of sentimentalism is here to stay because what it does, again and again, is correlate body with mind in convincing social contexts—and we can never get enough of such correlations, greedy mind readers that we are.[36]

Now think of the effects of claiming that sentimentalism is an eighteenth-century phenomenon and that "we, the modern public, no longer find it in ourselves to be moved" by *La Piété filiale* the way Greuze's contemporaries did. On the one hand common sense suggests that this claim is correct. Surely in the 1760s they must have responded to *La Piété filiale* somewhat differently than we respond to it now, just as audiences in the 1960s must have responded to Bobby Vinton's "Roses Are Red (My Love)" somewhat differently than we may now:

Roses are red, my love.
Violets are blue.
Sugar is sweet, my love,
But not as sweet as you.[37]

On the other hand, one practical effect of this claim is that sentimentalism begins to seem safely *contained*—sealed off as a relic of a long-gone epoch associated with a very specific list of texts and works of art. And, so contained, sentimentalism becomes usable again. That is, whatever writers and artists do now can be sentimental, but it cannot add up to "sentimentalism," for we have been done with that for more than two hundred years, haven't we?

And such containment and recycling are necessary, given that authors are always in need of new ways to render the body convincingly transparent. The rubric sentimentalism covers a broad variety of representational methods, many of which can never really go out of use. In fact, we can say that, when one method of forcing the body into transparency is declared passé and appended with a proper condescending "ism," it is an indication that this method is now on the way to being recycled in a different guise and reinvented by a new genre or group of artists.

Why We Like Absorptive Paintings

Throughout *Absorption and Theatricality* Fried remains interested in the interplay between psychology and history. Here, once again, though not a cognitive scientist himself—not even a fellow traveler—what he says is very compatible with the cognitive theory of mind reading. Indeed, his argument is better understood if we look at it squarely from a cognitive perspective.

Fried begins with a strong assertion of the historical limits of his argument. "This study is exclusively concerned with developments in France," he tells us on the first page. Then again on page 2: "I am convinced that there took place in French painting starting around the middle of the century a unique and very largely autonomous evolution; and it is the task of comprehending that evolution as nearly as possible in its own terms—of laying bare the issues crucially at stake in it—that is undertaken in the pages that follow."

By insisting that the French absorptive paintings should be considered on their "own terms," Fried wants to distance himself from several interpretive traditions.[38] Specifically, he disagrees with those art historians who think that by focusing on the human body in action, Chardin and others took an anti-rococo stance, reacting as it were against rococo's emphasis on decorative elements and its indifference to historical figures and heroic endeavors. As Fried sees it, authors of absorptive paintings were not really interested in upholding "the doctrines of the hierarchy of genres and the supremacy of history painting as they were held by anti-Rococo critics and theorists" (75). In his view the artists' interest in the representation of absorption was not ideological or primarily concerned with the subject matter. Instead it was "determined by other, ontologically prior concerns and imperatives" (75). And these had to do, among other things, with the relationship, "at once literal and fictive, between painting and beholder" (76).

Such "ontologically prior concerns"—particularly when framed in terms of the relationship "between painting and beholder"—are bound up with the cognition of mind reading. The absorptive painting titillates us with embodied transparency. Our responses to this powerful illusion of direct access certainly draw on an idiosyncratic mix of personal ideologies and aesthetics, but the sociocognitive—the drive to read minds and the anxiety about misreading minds—is inextricably there, heightening and shaping our interest in the painting.

In other words if we approach Fried's arguments squarely from the point of view of cognitive theory, we gain a better understanding of the intrinsic pull of absorptive paintings. Fried notes that "absorption emerges as good in and of itself, without regard to its occasion" (51). We can now say that a representation of absorption may feel "good in and of itself" because it flatters our mind-reading adaptations. Such representations titillate us with visions of perfect access to other people's minds, and they intensify our pleasure by constructing plausible social contexts for these fleeting mind-reading feasts.

So *still* without bridging the gap between mind reading and aesthetics in visual art, we can say that sociocognitive satisfaction may underlie aesthetic pleasure. It does not *define* this pleasure: too many culture-specific and personal idiosyncratic factors are at play in each case. In fact, as Fried demonstrates, a number of eighteenth-century critics found various faults with absorptive pieces, which means that a visual depiction of privileged mind access does not directly translate into a publicly acknowledged aesthetic pleasure for everyone. Still, at least to some degree, it makes this pleasure possible.

TEN

‖‖‖

{
In which women first appear more inscrutable than men, but then men catch up, and everybody is mysterious all around.
}

Painting Mysteries

Proposal Compositions

To portray somebody completely absorbed in what they are doing is an effective way to make their thoughts seem transparent to us, but it's not the only way. I turn now to another pictorial tradition that achieves the same effect using a very different strategy. As in the case with Fried, I rely on the work of a cultural historian, himself remote from cognitive science, whose analysis is nevertheless compatible with insights about the workings of our theory of mind.

Stephen Kern's *Eyes of Love* identifies a striking pattern in European genre painting of the second part of the nineteenth century. When "French and English artists . . . depicted a man and a woman in the same composition [they] typically rendered the face and eyes of the woman with greater detail and in more light. Most important, the men are in profile, while the women are frontal." These works exemplify what Kern calls a "proposal composition":

Such a composition highlights the woman's moment of decision after the man has proposed that the relationship move to some higher level of intimacy. At such moments she must respond, whether it be to his searching look or friendly inquiry, or more significantly, to his seductive offer or proposal of marriage. Her eyes convey an impending answer to the question *Will she or won't she?* And because she is thinking about many possible consequences of her answer, her expression is especially intriguing. In contrast, the man has done his thinking and said what is on his mind. He wants to hear a *Yes*, so his face bears a more predictable and less interesting expression.[1]

Kern makes a convincing case against the accepted critical view that visual representations of women—especially beautiful women—always objectify them. He argues that in proposal compositions, such as, for example, William Henry Midwood's *At the Crofter's Wheel* (fig. 34), women "are not objectified by the male gaze but retain a commanding subjectivity that, in comparison to the man's more erotically focused purpose and expression, conveys a wider range of thoughts and emotions" (228).

More interesting, *less* predictable, *especially* intriguing, conveying a *wider* range of emotions—this is our rule of contrasts at work. Contrasts and degrees are at the heart of proposal compositions. Using the familiar social script of courtship, proposal compositions construct the context in which the body of one protagonist is maneuvered into embodied transparency. We know what the man is thinking, his "erotically focused" purpose made even more obvious because it is contrasted with the "intriguing" thought processes of the woman.

The rule of transience is also present. We know what the man is thinking, but this moment of transparency cannot last. Depending on the woman's reaction, the man will soon adopt a different posture, attempting perhaps to conceal his disappointment if she says *no* or hesitates for too long. The same cultural narrative—the courtship narrative—that makes the instance of transparency convincing ensures that it is but an instance, serendipitously "caught" by the artist.

Something else might be at work in sustaining this illusion of a serendipitously caught moment of transparency: it steals upon the spectator unexpectedly. After all, "proposal compositions" were not known

FIGURE 34. William Henry Midwood, *At the Crofter's Wheel*, 1876. Oil on canvas, 28 × 36 in.

as such by their contemporaries. "Proposal composition" is a term introduced by Kern to describe not a particular subgenre associated with a specific style but a recurrent compositional pattern and interpersonal dynamic that can be found across different schools, styles, and representational traditions of the second part of the nineteenth century.

Contributing to this effect of unexpectedness are the titles of proposal compositions. They rarely indicate that we are witnessing a scene of romantic inquiry and hesitation. Very few titles are leading, such as *The Proposal, Pleading,* or *Showing a Preference.* The majority are all over the place, referring to a setting, to a prominent artifact present on the scene, or to the protagonist: *A Dance in the Country, Nameless and Friendless, Waiting for the Ferry, Effie Dean, A Rest by the Seine, Blossom Time, The Picnic, The Umbrellas, At the Crofter's Wheel.*[2] Thus, because there is neither an explicit genre affiliation nor a title that would mark a proposal composition as such, viewers have to infer on their own, in the case of each specific painting, that it contains a deliberation induced by a proposal of marriage or romantic liaison.

So as contemporary spectators approached a painting, say, Renoir's *A Dance in the Country,* Osborn's *Nameless and Friendless,* or Horsley's *Blossom Time,* they did not know beforehand that the body of one of its protagonists was supposed to be strikingly readable (even though, once they started looking at it, they could see it right away).[3] Had it been known—that is, had the proposal composition indeed emerged as an established subgenre with its own set of typical titles—the man's transparency would have eventually become a convention and as such would have required extra effort to be rendered convincing (as seemed to have happened with eighteenth-century absorptive paintings). This did not happen, however: the impression of serendipity was not marred by the thought that in this artistic subgenre serendipity *is* a convention.

I am not making any teleological claims about the history of proposal compositions. That is, I am not saying that late nineteenth-century artists and art critics consciously avoided recognizing a new subgenre in their midst in order to continue crafting compelling narratives of embodied (male) transparency. It seems, rather, that authors of proposal compositions differed widely from each other in their styles and sensibilities and did not give much thought to this particular common denominator, and neither did their audiences. We needed Kern's book to finally see this common denominator, and we need research on theory of mind to see why we intuitively value the moments when bodies reveal minds so vividly.

Problem Pictures

While I was writing this book, I gave several talks about proposal compositions as an example of embodied transparency. My audiences were typically of two types: cognitive scientists and literary scholars. Afterward, both asked me an interestingly different version of the same question. Cognitive psychologists asked, Is it possible that male artists generally tend to depict women as more mysterious and that the proposal compositions reflect this tendency? Literary critics asked, Is it possible that, given the late Victorian period's anxiety about female sexuality, contemporary artists did indeed portray women's thought processes as particularly intriguing?

In response to these questions I now turn briefly to another tradition in genre painting, one that partially overlapped with proposal compositions in time, was largely represented by male artists, and frequently focused on a conversation between a man and a woman. Although such paintings, known as "problem pictures," are similar to proposal compositions, they neither depict women as more mysterious than men nor feature embodied transparency.

To repeat: I am using problem pictures here to argue *against* the view that the reason proposal compositions portray women as more inscrutable is the cultural anxiety, shared both by artists and their audiences, about women's intentionality. (Though I certainly agree that this anxiety plays *some* role, I will address this issue later.) I believe that what is at stake specifically in proposal compositions—what makes them special and different from other genre paintings—is their emphasis on embodied transparency. Hence, they portray women as more mysterious not because this reflects some general feeling about how women are to be portrayed but because by doing so they can make men seem much more readable. The women's opacity, in other words, serves to create the contrast necessary for the construction of embodied transparency in men, and the cultural context of courtship makes this contrast socially plausible.

What are problem pictures? According to Pamela M. Fletcher's study *Narrating Modernity* they "were an extraordinarily popular feature of the Edwardian Royal Academy. The term referred to ambiguous, and often slightly risqué, paintings of modern life which invited multiple, equally plausible interpretations" (1). Consider John Collier's *A Confession* (1902), which depicts a "couple engaged in an emotional conversation," in which the woman's face is "in shadow, while the man, leaning forward with his elbows on his knees, brings his face into the light of the fire,"[4] staring down and slightly to the viewer's right (fig. 35).

Drawn to the couple's feelings yet unable to read them, the visitors to the academy turned their mind-reading energies to their own and the artist's mental states. According to Collier he received "many inquiries"

FIGURE 35. John Collier, *A Confession*, 1902. Oil on canvas, 44 × 56 in.

about the picture's subject. In one extant letter the "writer pleads: '*Oh! Honourable John, I want to know very badly which (please tell me) is confessing in your Royal Academy picture—the man or the woman.*'" In response to such queries the artist offered "multiple interpretations," cultivating "oracular ambiguity" along the lines of "the woman did it and the man confessed it."[5] In 1913 he revisited the subject with his *Fallen Idol*, in which a "woman kneels at a man's feet, her upper body resting on his knees and her head bowed in an attitude of grief or shame. The man holds one of her hands in his, and stares directly out of the canvas, his face illuminated by a shaft of light."[6] As Fletcher reports,

> Critical responses in the press were almost equally divided between those who read the story as completely open to interpretation, and those who assumed that the woman had "fallen." The *Daily Mir-*

ror, Queen, Reynolds's, and the *Daily Sketch* all read the picture as ambiguous, predicting, "Lots of stories will be woven around this picture, and probably none of them the right one." The *Daily Mirror* made the point by "quoting" two visitors: "'Of course, he's just confessed something he's done,' said one woman yesterday confidently. 'She's just been found out,' said the next comer with equal assurance." (130–31)

The spatial arrangement of the figures and the pattern of lighting do not consistently single out one sex as more mysterious than the other: the man is just as likely as the woman to be "in the position of the 'problem.'"[7] Also, time is not an issue. For instance, in *A Confession* there is no indication how long the man and the woman have been in their present positions or how long they may continue in them. Because the artist's goal is "antitransparency" (if I may put it this way), neither the rule of contrasts nor the rule of transience applies.

If the earlier discussion of absorptive paintings has made you wonder whether *any* subject who is not aware of the beholder is transparent, the problem pictures answer this question in the negative. Neither the man nor the woman in *A Confession* seems to be conscious of any beholder, but they are certainly not transparent.

Yet they *are* absorbed in their thoughts—which leads me to suggest that, even though the majority of absorptive paintings depict transparent subjects, there are some exceptions to this rule. If an artist succeeds in portraying a subject who is completely absorbed even as the object of her absorption remains a mystery to us—which, I suspect, is rather difficult—then we get absorption without transparency (and turn our mindreading efforts toward ourselves and the artist).

Fletcher demonstrates that interpretations of problem pictures reflected some of the "most pressing issues of the early twentieth century, including the nature of modern marriage and motherhood, the emergence and definition of the new professional classes, and the existence of a specifically feminine morality" (1). No doubt many of the same issues were in play in proposal compositions, but observe the crucial difference between their respective constructions of protagonists' subjectivity. A problem picture leaves its subjects' feelings largely open to interpretation and only somewhat constrains them within broad categories, such as "distress" or "surprise." In contrast, a proposal composition constructs one participant as decisively more readable.

Fletcher's emphasis on the cultural surround of problem pictures returns us to the earlier question of whether proposal compositions responded to *their* cultural surround, reflecting the late nineteenth-century anxiety about female sexuality. Now we can respond to this question with a yes, but a yes qualified by what we know about our mind-reading adaptations.

In principle, every period in human history is characterized by an anxiety about female sexuality. It is but another component of our broader mind-reading uncertainty. Female bodies do not advertise their sexual intentions. To use just one example, the concealed estrus makes it impossible for men to be certain about paternity, which means that they have to second-guess their partners' intentions of staying faithful to them.[8] The endeavor to control female sexuality—which may take different forms in different cultures—is thus really the endeavor to control women's intentions and thus to minimize this particular (i.e., related to paternity) aspect of mind-reading uncertainty.

But this also means that a hypothesis that a given group of paintings featuring women reflects its time's anxiety about women's sexuality will be true about *any* group of paintings featuring women. It is thus *trivially* true because it does not predict anything about any specific painting or representational tradition. It cannot explain, for example, why in proposal compositions women are portrayed as more mysterious than men while in problem pictures they are clearly not.

To explain this difference, we have to turn to cognitive science and suggest that authors of proposal compositions and problem pictures have different goals in representing the thought processes of their subjects. Both are obsessed with mind reading—all paintings depicting people are. But authors of proposal compositions emphasize that we have unequal access to the minds of their protagonists, which results in a state of embodied transparency of one protagonist but not another. (Not that they think in these terms, of course.) The mind-reading mode of problem pictures is very different. The artists set themselves the challenge of creating a social context in which we have very little access, and this lack of access keeps us enthralled by the painting, registering our own puzzlement about its subjects' mental states and trying to guess the artist's intentions.

I just used the word *enthralled* on purpose. I did it to echo the rhetoric of Fried in *Absorption and Theatricality*, in which he talks about the "enthrallment" of the viewer by an absorptive painting. I want to stress that there is no one correct or better way for a painting to engage our theory of mind and keep us riveted. Different paintings can do it differently; in fact, they can use diametrically opposite techniques to enthrall their spectators.

Absorptive paintings and proposal compositions do it by cultivating embodied transparency—by making us think that we know exactly what some of their subjects think. Problem pictures do it by focusing us on the subjects' state of mind yet preventing us from figuring it out, thus sending our theory of mind racing in three directions simultaneously ("What are they thinking?" "I don't get it!" "What did the artist have in mind?"). We could look at a variety of other traditions and see how they experiment with the balance between information about their subjects' mental states that they give us outright and that they want us to keep guessing at, thereby turning us toward our own and the artist's thought processes.

And as we think about our thinking about a work of art, we may come to revise our understanding of what constitutes a particular representational tradition. For just as there is no predicting what cultural forms mind reading will assume at a given historical moment, so there is no predicting what forms thinking about these forms will assume. A culture of greedy mind readers never ceases to supply and demand endlessly mutating, endlessly nuanced, endlessly new configurations of mental states.

Coda

- Theory of mind evolved to track mental states involved in real-life social interactions.
- On some level, however, our theory-of-mind adaptations do not distinguish between the mental states of real people and of fictional characters.
- Cultural representations, such as novels, drama, movies, paintings, and reality shows, indulge our greedy theory of mind, giving us carefully crafted, emotionally and aesthetically compelling social contexts shot through with mind-reading opportunities.
- Hence the pleasure afforded by following minds on page, screen, stage, and canvas is to a significant degree a *social* pleasure. It's an illusory but satisfying confirmation that we remain competent players in the social game that is our life.
- One of numerous strategies used by cultural representations to intensify this pleasure is to present our mind-reading adaptations with fantasies of embodied transparency, that is, with complex social contexts in which people's bodies seem to provide direct access to their minds.

- Embodied transparency of this kind is rare in real life, in which our perception of direct access to a person's mind is usually inversely related to the complexity of the social situation at hand.
- Although there is no predicting what forms the fantasy of embodied transparency may take at a specific cultural moment, certain patterns—such as transience, contrasts, and restraint—seem to recur in its representation. These patterns hold more sway in some genres than in others; for instance, transience is more important for novels and paintings, restraint for movies.
- As soon as a culture becomes aware of an established niche for representing embodied transparency, this niche is vulnerable to subversion and parody. Hence writers, artists, and, more recently, film directors and television producers are always on the lookout for new convincing ways to portray bodies as providing direct access to minds.

Notes

IIIIIIIIIIIIIIIIIIIIIIIIIIIIIIIIIIIIIIIIIIIIIIIIIIIIIIIIIIIIIIIIIIIIIIIIIIIIIIIIIIIIIIIIIIIIIIIIIIIIIIIIIIIIIIIIIIIIIIIIIIIIIIIIIIIIIIIIIIIIII

Preface

1. For more information about the field of cognitive cultural studies, see Zunshine, "What is Cognitive Cultural Studies."

One: Culture of Greedy Mind Readers

1. Baron-Cohen, *Mindblindness,* 21. For foundational works on theory of mind as well as important recent studies, see Byrne and Whiten, *Machiavellian Intelligence* and "The Emergence of Metarepresentation"; Dunbar, "Evolutionary Basis of the Social Brain"; Gomez, "Visual Behavior"; Frith and Frith, "Social Cognition in Humans"; Keenan at al, "An Overview of Self-Awareness and the Brain"; Nettle, "Emphasizing and Systemizing" and "Psychological Profiles;" Premack and Dasser, "Perceptual Origins"; Saxe, "Why and How"; Saxe and Kanwisher, "People Thinking about Thinking People"; and Stiller and Dunbar, "Perspective-Taking." For a discussion of alternatives to the theory of mind approach see Dennett, *The Intentional Stance.*

2. See Tooby and Cosmides, "The Psychological Foundations of Culture," in Barkow et al., *The Adapted Mind.*

3. For a critique of the term theory of mind as it is used in the social neuroscience literature, see Stone and Hynes, "Real-World Consequences of Social Deficits," 462.

4. Borenstein and Ruppin, "The Evolution of Imitation," 229.

5. Rizzolatti et al., "Neuropsychological Mechanisms," 662.

6. See Bloom, *Descartes' Baby* (113–15), for a discussion of mirror neurons in the context of empathy and compassion. For a discussion of the relationship between mirror neurons and theory of mind in the context of simulation theory see Goldman, *Simulating Minds.*

7. Keysers et al., "The Mirror Neuron System and Social Cognition," 530–31.

8. Note, for instance, that, as Jochen Triesch and his colleagues observe, the evidence for mirror neurons in humans is indirect. Because "direct observation of [human] individual mirror neurons is impossible with today's experimental techniques," cognitive scientists have to infer it via converging data from "functional magnetic

resonance imaging (fMRI), electroencephalography (EEG), and magnetoencephalography (MEG) studies." Moreover, "nothing is known about what experiences and interactions with the environment are necessary or sufficient for the emergence of mirror neurons" ("Emergence of Mirror Neurons," 150–51). See also Triesch et al. for a discussion of recent experiments investigating the "question of whether mirror neurons are innate or whether they acquire their special properties through a learning process" (161); Hickok, "Eight Problems for the Mirror Neuron Theory of Action Understanding in Monkeys and Humans"; and Bauman et al., "The Neurobiology of Primate Social Behavior," 692–93.

9. Cognitive scientists thus begin to enter territory that has been extensively charted by philosophers and literary critics exploring mimesis (from Aristotle's *Poetics*, David Hume's "Of Tragedy," Erich Auerbach's *Mimesis*, and Walter Kaufmann's *Tragedy and Philosophy* to the recent rethinking of mimesis and performativity in cultural studies), phenomenology (such as George Butte's compelling reintroduction of Maurice Merleau-Ponty into literary and film studies, *I Know That You Know That I Know*), and intentionality (such as Martha Nussbaum's critique of the tradition of correlating "an emotion and a discernible physical state" [*Upheavals of Thought*, 96]). Although the work on mirror neurons is still in a relatively early stage, one can see exciting possibilities emerging at the intersection of traditionally humanistic research and the inquiry into the neural basis of interpersonal subjectivity.

10. Priborkin, "Cross-Cultural Mind Reading," n.p.

11. See Csibra, "Goal Attribution to Inanimate Agents"; Luo and Baillargeon, "Can a Self-Propelled Box Have a Goal?"; Song and Baillargeon, "Infants' Reasoning"; Song et al., "Can an Actor's False Belief Be Corrected"; and Baillargeon et al., "The Development of False-Belief Understanding."

12. Shany-Ur and Shamay-Tsoory, "Theory of Mind Deficits," 936.

13. Blair, "Theory of Mind, Autism, and Emotional Intelligence," 419. Though, as Ralph Savarese reminds us, the ability to follow social rules depends also on adequate control over one's nervous system. As he puts it, "Which is decisive for proper comportment: awareness of those rules or having a body and a nervous system that allow actually following them? Someone with Tourette's, for example, perfectly understands the inappropriateness of shouting "Fuck" at church but cannot stop himself from doing so" (personal communication, January 9, 2012).

14. Kleiner, *Gardner's Art through the Ages*, 277.

15. See Kelly et al., "Social Experience."

16. Schultz, "Developmental Deficits," 125. Interestingly, there seems to be a developmental shift in face-reading by infants between the ages of seven and ten months: "7-month-olds discriminate between facial expressions based on feature information rather than on affective meaning. On the other hand, older infants (10 months) are able to identify common affect among facial expressions and discriminate them from novel expressions. Additionally, it was shown that infants can use others' angry and happy facial cues to disambiguate uncertain situations and regulate their behavior ac-

cordingly" (Grossman et al., "Developmental Changes," 35). For further discussion see Sorce et al., "Maternal Emotional Signaling."

17. See Guthrie, *Faces in the Clouds.*

18. See Zebrowitz, *Reading Faces;* Zebrowitz and Zhang, "The Origins of First Impressions"; and Todorov et al., "Understanding Evaluation of Faces."

19. See Arendt, *The Life of the Mind:* "From the very onset, in formal philosophy, thinking has been thought of in terms of *seeing.* . . . The predominance of sight is so deeply embedded in Greek speech, and therefore in our conceptual language, that we seldom find any consideration bestowed on it, as though it belonged among things too obvious to be noticed" (110–11). Compare this to David Michael Levin's observation, drawing on Hans Jonas's essay "The Nobility of Sight," that "from the very dawn of our culture [sight] has been thought to be the noblest of the senses" ("Introduction," 2).

20. See Baron-Cohen, *Mindblindness,* for a discussion of mind reading in congenitally blind people.

21. See Bering, "The Existential Theory of Mind," 12.

22. For a brilliant discussion of how and why we attribute mental states to people who are dead, see Jesse Bering's recent book, *The Belief Instinct.*

23. See Dutton, *The Art Instinct,* for a response to critics who think that people from "other cultures" don't have what we call art, particularly the part "But They Don't Have Our Concept of Art."

24. See, for example, Alex Pentland's *Honest Signals,* in which he suggests that people become convinced not by the strength of an argument but by the relative evenness of affect displayed by the person who is trying to convince them of something.

25. Goffman, *Strategic Interaction,* 80–81.

26. Theater historian Joseph Roach, for instance, has argued that performance, "though it frequently makes references to theatricality as the most fecund metaphor for the social dimensions of social production, embraces a much wider range of human behaviors. Such behaviors may include what Michel de Certeau calls 'the practice of everyday life,' in which the role of spectator expands into that of participant" (Roach, "Culture and Performance,"46).

27. Our everyday mind reading turns each of us into a performer and a spectator, whether we are aware of it or not. As Ellen Spolsky argues in "Narrative as Nourishment": "The clues to which we sensibly learn to be attentive cannot be relied on absolutely because bodies themselves, the bodies that are evolved to give external expression to internal states, learn to produce these clues within contexts differentiated by cultural categories such as gender, age, social class, and occupation. Not only our interpretations of them but the evolved physical expressions themselves are enriched and/or distorted by social overlays, making both misinterpretation and deliberate deception possible" (48–49). Thus, a particular body can be viewed as a time-and-place-specific cultural construction, that is, as an attempt to influence others into perceiving it in a certain way. Compare this to Hegel's argument about the insta-

bility of the inner mental state when it is made visible: "The inner in thus appearing is doubtless an invisible made visible, but without being itself united to this appearance. It can just as well make use of some other appearance as another inner can adopt the same appearance. Lichtenberg, therefore, is right in saying: 'Suppose the physiogno-mist ever did have a man in his grasp, it would merely require a courageous resolution on the man's part to make himself again incomprehensible for centuries' " (Hegel, *The Phenomenology of Mind,* 345).

28. As Margaret Talbot observes: "Maybe it's because we're such poor lie de-tectors that we have kept alive the dream of a foolproof lie-detecting machine. This February, at a conference on deception research, in Cambridge, Massachusetts, Steven Hyman, a psychiatrist and provost of Harvard, spoke of 'the incredible hunger to have some test that separates truth from deception—in some sense, the science be damned' " ("Duped," 54).

29. Again, compare this cognitive-evolutionary insight with the work done by cultural theorists ranging from Judith Butler to Peggy Phelan, who have written exten-sively on the body as a constantly receding signified, a perennially contested deposi-tory of reliable meanings. Think, for example, of Phelan's observation that whereas "the living performing body is the center of semiotic crossings, which allows one to perceive, interpret and document the performance event," we long to "return to some place where language is not needed," an "Imaginary Paradise" where there are no "linguistic and visual distinctions between who one is and what one sees" ("Reciting the Citation of Others," 15, 29). Think, too, that some of the resistance to the view of the body as always constructed and always performed can come from our hoping against all hope that it *must* be possible to carve some zones of certainty in the exas-perating world where our favorite source of information, the body, is often untrust-worthy in direct proportion to the extent to which we trust it.

30. As Catherine Gallagher and Stephen Greenblatt put it in *Practicing New Historicism,* the body always "functions as a kind of 'spoiler' . . . baffling or exceeding the ways in which it is represented" (15).

31. And what you yourself are thinking, too (see Palmer, *Fictional Minds*).

32. Compare this to Baron-Cohen's description of Daniel Dennett's view of the "intentional stance": "Dennett is not committed either way on the question of whether there really are such things as mental states inside the heads of organisms. We ascribe these simply because doing so allows us to treat other organisms as rational agents" (Baron-Cohen, *Mindblindness,* 24).

33. I concede that in a sci-fi movie this jump may mean that there was a magnet planted in the person's body and that he was pulled up by aliens who use such magnets to reel in earthlings to their ship. Note two things, though. To counterbalance our immediate tendency to read a mental state into his behavior, I had to come up with a truly outlandish scenario. Theory-of-mind-free explanations apparently require quite a bit of work. And, furthermore, my explanation is not really completely theory-of-mind-free. I didn't manage to get rid of intentionality altogether. Only instead of ascribing an intention to the man, I ascribed it to aliens.

34. Spolsky, *Satisfying Skepticism,* 7; and Spolsky, "Darwin and Derrida," 52.

35. Sperber, *Explaining Culture,* 38.

36. At least not initially. And, yes, I am speaking of Andy Kaufman.

Two: I Know What You're Thinking, Mr. Darcy!

1. Austen, *Persuasion,* 66–67.

2. For an important broader discussion of the effect of cultural representations on our evolved cognitive adaptations see Bloom, *How Pleasure Works.*

3. The centrality of mind-reading to our enjoyment of fictional narratives continues to be borne out by ongoing studies in cognitive psychology (e.g., Barnes, "Fiction and Empathy" and Barnes et al., "Reading Preferences"), although such studies would benefit from greater input from autistic individuals themselves, especially those labeled as "low-functioning" and who have learned to communicate by typing. For a discussion, see Savarese and Savarese, *Autism and the Concept of Neurodiversity.*

4. As Alan Palmer puts it, "Novel-reading is mind reading. Fiction can only be understood in this way" ("Storyworlds and Groups," 182).

5. Having considered this view at length elsewhere to be a case of mild professional hypocrisy (see my "Cognitive Alternatives to Interiority"), I only want to point out here that we can't make sense of fictional worlds if we don't allow characters the same capacity for mental states as we do the people who surround us. Like it or not, we perceive fictional characters as having theory of mind and thus can't avoid speculating about their mental states in this or that hypothetical situation just as we do about the mental states of real people.

6. Taking a cue from the essayist Phillip Lopate, one can further expand the list of specifically sexual expressions of embodied transparency. As Lopate observes, for him "[the] power of the flaccid penis's statement, 'I don't want you,' is so stark, so cruelly direct, that it continues to exert a fascination out of all proportion to its actual incidence" ("Portrait of My Body," 333).

7. See Dennett, *Consciousness Explained,* 107.

8. Kurzban, *Why Everyone (Else) Is a Hypocrite,* 44, 21.

9. Compare this to Ellen Spolsky's argument in *Gaps in Nature,* a book strikingly ahead of its time.

10. Kurzban, *Why Everyone (Else) Is a Hypocrite,* 5. For an important related argument see Carruthers, *The Opacity of Mind.*

11. For a discussion of essentialist thinking see Zunshine, *Strange Concepts.*

12. Sure, an author may claim to have no clue what her or his characters are thinking, but this is a kind of claim we are happy to believe in abstract, until we are faced with a concrete mind-reading mystery. For instance, do you think Nabokov really didn't know if Charles Kinbote dreamt Zembla up or if it actually exists? He said he didn't know, but I don't believe him. As the philosopher Colin McGinn puts it compellingly, "Fictional characters have the mental states they are represented as having; and the basic reason for this is that fictional characters have the characteristics,

mental or physical, they are said by their authors to have. Real people, by contrast, may not have the mental states they purport to have; it is always a mere *hypothesis* that they have the mental states you think they have (*The Power of Movies,* 122).

13. For a detailed discussion see Zunshine, *Why We Read Fiction,* esp. 71–72.

14. For a discussion of Barthes's and Foucault's "Death of the Author" concept in the context of theory of mind see Zunshine, *Why We Read Fiction,* 66–67.

15. Compare to Sternberg's observation about the writer's "godlike privileges of unhampered vision, penetration to the innermost recesses of the [fictional] agents' minds, free movement in time and space, and knowledge of past and future" (*Expositional Modes,* 257). See also Yu, *Rereading the Stone,* 166; and Booth, *The Rhetoric of Fiction,* 3.

16. Gombrich, *Art and Illusion,* 298.

17. For a suggestive argument about Jane Austen's use of contrasts see Woloch's *The One vs. the Many,* chap. 1, "Narrative Asymmetry in *Pride and Prejudice.*"

18. Fridlund, "Evolution and Facial Action," 30, 21, 37. See also Seyfarth and Cheney, "Signalers and Receivers"; Russell et al., "Facial and Vocal Expressions of Emotion"; Barrett et al., "On the Automaticity of Emotion"; Ekman, "Strong Evidence"; and Ekman and Fridlund, "Assessment of Facial Behavior in Affective Disorders."

19. See Zunshine, "1700–1775"; and Zunshine, "What to Expect."

20. Austen, *Pride and Prejudice,* 129.

21. Compare this to Cohn's discussion in *Transparent Minds* of E. T. A. Hoffmann's *Master Flea.* Cohn writes that in Hoffmann's story "the microscopic magician of the title gives to his human friend Peregrinus Tuss a tiny magic lens, that, when inserted in the pupil of his eye, enables him to peer through the skulls of all fellow human beings he encounters, and to discern their hidden thoughts. Peregrinus soon curses this 'indestructible glass' for giving him an intelligence that rightfully belongs only to the 'eternal being who sees through to man's innermost self because he rules it' " (3). Peregrinus thus forces "all fellow human beings" into a state of embodied transparency *that can last infinitely*—an ethically indefensible situation that is resolved to the extent to which Peregrinus is rendered unhappy by his privileged access. For a related argument see Spolsky, "Elaborated Knowledge."

22. Tiedens, "Anger and Advancement" (quoted in Butler and Gross, "Hiding Feelings," 114).

23. Fielding, *Bridget Jones,* 73.

24. Hellman and O'Gorman, *Fabliaux,* 111.

25. Vitz, "Tales with Guts," 157–58.

26. Fielding, *Tom Jones,* 62, 63.

27. Of course, one may suggest that Allworthy is smarter than he appears and that he is, in fact, at least partially aware of the doctor's double game. This is an appealing reading, but we have no evidence for it in the text.

28. Palmer, *Fictional Minds,* 10; see also McGinn, *The Power of Movies,* 122.

29. Tolstoy, *Anna Karenina*, 335.

30. Defoe, *Robinson Crusoe*, 90.

31. Austen, *Emma*, 123.

32. Hemingway, *A Farewell to Arms*, 12. For a brief related discussion of theory of mind and Hemingway's style see Zunshine, *Why We Read Fiction*, 23.

Three: Sadistic Benefactors

1. Michaels, *The Shape of the Signifier*, 70.

2. For an intriguing related argument about theory of mind and fictional depictions of "excessively cruel people" see Vermeule, *Why Do We Care?* 86.

3. Fielding, *The History of Ophelia*, 1:252.

4. See Wendy Jones for a fascinating discussion of cultural differences underlying potentially conflicting perspectives on what constitutes philanthropy. As she puts it, "while sympathy in its technical sense as the recognition of emotion is a feature of cognition, cultures and individuals have developed countless ways to block or ignore sympathetic response: the torturer and the philanthropist will view pain very differently" ("*Emma,* Gender," 332)

5. Rousseau, *Emile*, 442.

6. Palahniuk, *Fight Club*, 155.

7. Using James Phelan's categorization of unreliable narrators, we can say that Tyler underreports and underreads (219) Raymond's mental states.

8. For this and many other useful suggestions I am grateful to the anonymous reviewer for the Johns Hopkins University Press.

9. It's almost the exact opposite of the retroactive reading of body language in detective stories. There, we revisit facial expressions and gestures that were initially ambiguous and find out what they really meant. Here, we revisit facial expressions and gestures whose meaning we thought we knew only to realize that we will never know for certain what the characters really felt on those occasions. Detective stories begin with ambiguity and end with transparency, and unreliable-narrator stories begin with transparency and end with ambiguity.

10. Compare to Foucault's argument in *Discipline and Punish* about the asymmetry of access implied by the respective positions of the prisoners and the guard in the Panopticon.

Four: Theaters, Hippodromes, and Other Mousetraps

1. Richardson, *Clarissa*, 640.

2. Hornby, *How to Be Good*, 69.

3. Richardson, *Clarissa*, 620.

4. Ibid., 99.

5. Siddons, *Practical Illustrations of Rhetorical Gesture and Action*, 35–36. Siddons's book is an 1807 adaptation of Johann Jacob Engel's earlier treatise *Ideen zu*

einer Mimik (1785). Part acting manual, part philosophical meditation on theater, it occasionally reads as a work of fiction, particularly when Siddons imagines various social situations involving performers and spectators.

6. Tolstoy, *Anna Karenina,* 209–10.

7. Hornby, *Fever Pitch,* 19–21.

8. The British movie version was made in 1997. There is also an American version made in 2005 that makes the protagonist a fan of the Boston Red Sox. The American *Fever Pitch,* however, does not foreground embodied transparency in any significant way.

9. Absorption is a concept highly relevant to discussions of transparency, as my forthcoming chapter on paintings will show.

Five: Movies

1. For recent and forthcoming investigations of literature, film, and the arts from the point of view of cognitive theory see Abbott, *The Fine Art of Failure, "Conversion," and "Reading Intended Meaning"*; Aldama, "Race, Cognition, and Emotion" and *Toward a Cognitive Theory of Narrative Acts;* Anderson and Anderson, *Moving Image Theory;* Austin, *Useful Fictions;* Bortolussi and Dixon, *Psychonarratology;* Branigan, *Projecting a Camera;* Carroll, *The Philosophy of Motion Pictures;* Crane, "Surface, Depth, and the Spatial Imaginary"; Currie, *Image and Mind;* Easterlin, *A Biocultural Approach to Literary Theory and Interpretation;* Flesch, *Comeuppance;* Fludernik, "1050–1500: Through a Glass Darkly"; Herman, *Story Logic, Narrative Theory and the Cognitive Sciences,* and "Genette Meets Vygotsky"; Hart, "The Epistemology of Cognitive Literary Studies"; Hogan, *Cognitive Science, Literature, and the Arts,* "Literary Universals," *Empire and Poetic Voice, The Mind and Its Stories,* and *Understanding Nationalism;* Tony Jackson, "Issues and Problems"; Keen, *Empathy and the Novel, Thomas Hardy's Brains,* and "Strategic Empathizing"; Kramnick, "Some Thoughts on Print Culture and the Emotions"; McConachie, *American Theater* and *Engaging Audiences;* Palmer, *Fictional Minds* and *Social Minds and the Novel;* Plantinga and Smith, *Passionate Views;* Richardson, *British Romanticism, The Neural Sublime,* "Studies in Literature and Cognition," and (with Ellen Spolsky) *The Work of Fiction;* Scarry, *Dreaming by the Book;* Spolsky, "Darwin and Derrida," "Narrative as Nourishment," *Gaps in Nature, Satisfying Skepticism,* and *Word vs Image;* Starr, *Feeling Beauty,* "Multisensory Imagery," and "Poetic Subjects"; Turner, *The Literary Mind;* Vermeule, "God Novels," "Satirical Mind Blindness," *The Party of Humanity,* and *Why Do We Care;* and Zunshine, *Why We Read Fiction, Strange Concepts,* and *Introduction to Cognitive Cultural Studies.*

2. McGinn, *The Power of Movies,* 104.

3. For a related discussion see the chapter on detective stories in my *Why We Read Fiction.*

4. Lane, "Miles to Go," 83.

5. For a brilliant analysis of gambling in fiction, from a different theory-of-mind perspective, see Vermeule, *Why Do We Care?*

6. Remnick, "Exit Bin Laden," 36.

7. Mazzetti et al., "Behind the Hunt for Bin Laden,"n.p.

8. Tapper, "Hillary Clinton Explains," n.p.

9. Quigley, "Maybe I Just Coughed," n.p. See also Gevorkian, "Ne povod dlia tanzev."

10. Austen, *Pride and Prejudice,* 130.

11. Tooby and Cosmides, "Consider the Source."

12. Pemberton, *The Hottest Water in Chicago,* 18, 13.

Six: Mockumentaries, Photography, and Stand-Up Comedy

1. Gervais and Merchant, *The Office: The Complete First Series,* disk 1. All references are to this edition.

2. Ibid., commentary.

3. Though, as Bloom notes in *How Pleasure Works,* "I know more than one person who finds it hard to watch comedies that rely too heavily on embarrassment; they find it almost unbearable" (166).

4. Here I use the terms *cinéma vérité* and *direct cinema* interchangeably, but there are important differences between the two. For an overview of these differences see Ellis and McLane, *A New History,* 216–18.

5. Quoted verbatim from Hope Ryden's official website, www.hoperyden.com/ disc.htm (accessed August 18, 2007).

6. Quoted in *Cinéma Vérité: Defining the Moment* (dir. Peter Wintonick).

7. Wintonick, *Cinéma Vérité.*

8. Ellis and McLane, *A New History,* 216, 219.

9. Ibid., 217. And I love the fact that the language here is so similar to Michael Fried's in his discourse on absorption (see Fried, *Absorption and Theatricality*).

10. Ellis and McLane, *A New History,* 217. The other side of this issue is that such revelatory moments were carefully staged and edited. As Frederick Wiseman, a pioneer of direct cinema, puts it, "It's all manipulation. Everything about that kind of movies is a distortion" (quoted in Wintonick, *Cinéma Vérité*). Wiseman refers to the manipulation of material done on *his* side of the camera, such as editing to intensify the drama, but other people have spoken of manipulation that occurs on the other side. Once the filmed subjects intuit that the value of the film consists in "capturing human emotion spontaneously as it happens," the whole endeavor becomes vulnerable to subversion. The protagonists may instinctively script their own moments of embodied transparency, and there is no knowing when they start and stop performing them.

A telling case in point is Albert and David Maysles's *Grey Gardens* (1976), a documentary that follows the daily routines of two Edith Beales, mother and daughter, former socialites, legendary beauties, and gifted singers, who now, aged eighty and fifty-seven, live in isolation in their dilapidated mansion in the East Hamptons. On the one hand Albert Maysles describes the goal of "direct cinema" as recording "life as it is—no better, no worse," the implicit premise here being that the emotions that we see on film arise naturally out of the course of everyday life, with moments of embodied transparency occurring spontaneously throughout. On the other hand the subjects of this particular film, both frustrated performers, whose careers onstage were cut short, thrive on living their emotions on camera. When they appear surprised, disappointed, or flustered, there is no telling how truly "transparent" they are. It is possible that they play up those moments of "direct access" for each other and their spectators.

Not accidentally, when later accused of taking the Beales's privacy "too far," Albert Maysles responded with the story that "Little Edie" told him about "Big Edie's" death. When, at her mother's deathbed, the daughter "asked her if she had anything more to say," Big Edie "said it was all on the film. It was the performance of a lifetime" (quoted in Peter Keough, "Shades of Grey"). You can also hear it in Albert Maysles's *The Beales of Grey Gardens* [2006]). In other words the Maysles were aware that the mother-daughter duo might be performing for their camera according to their own "scripts" and emotional needs.

11. My argument here is directly influenced by Michael Fried's *Absorption and Theatricality*—to be discussed in the last two chapters—specifically by his observation that Jacques-Louis David's repeated portrayals of Homer and Belisaire, both blind and hence presumably unable to perform their body language for observers, were driven by David's desperate search for new reliable contexts for absorption.

12. *The Broadcast Tapes of Dr. Peter* (1994) and *Silver Lake Life: The View from Here* (1990) were made by filmmakers diagnosed with AIDS, who chronicled their fight with the disease until they died. For a discussion of these movies see Ellis and McLane, *A New History*, 284–87.

13. For a discussion of the history of that moment see Ellis and McLane, *A New History*, 290–91.

14. See ibid., 236–37.

15. I am lifting this phrase and its meaning directly from Fried's *Absorption and Theatricality* (61).

16. Zehme, *Lost in the Funhouse*, 256.

17. Zmuda, *Andy Kaufman Revealed!* 253.

18. Ibid., 321.

Seven: Reality TV

1. I wonder if the first strategy is used more often with young female participants, who, in Western culture, are expected to express their emotions publicly and are forgiven for it, and the second, with older and male participants.

2. For further discussion see Zunshine, *Why We Read Fiction.*

3. Quoted at http://pajamasmedia.com/blog/confessions_of_a_reality_junki (accessed Oct. 21, 2011).

4. See, for instance, research on neurocognition of the human voice, according to

which "a speaker never produces the same sound twice" (Belin, "'Hearing Voices,'" 387).

Eight: Musicals

1. See, e.g., Dunne, *American Film,* 79.

2. McMillin, *The Musical as Drama,* 102–3. For an important related discussion of backstage musicals see Feuer, *The Hollywood Musical,* 5–14.

3. Although it could be argued that when Higgins sings "Why Can't the English," Pickering, Eliza, and the random passersby hear his impassioned speech but not the accompanying orchestra. In fact, what the orchestra is doing might be different yet, for it may "sing" its own melody, inaccessible to characters. For a discussion of how this happens in musicals see McMillin, *The Musical as Drama,* 130–45; for a related discussion of operas see Hutcheon and Hutcheon, "Narrativizing the End," 443.

4. As McMillin puts it, such songs "are not called for as numbers by the book but are forms of spontaneous expression by the characters" (112).

5. See Kurzban, *Why Everyone (Else) Is a Hypocrite.*

6. In the 1955 movie version, once Marlon Brando begins to sing, the lighting around him changes to emphasize that we are entering a different space and time.

7. For a discussion of different interpretations of what will happen next see Miller, *From Assassins to West Side Story,* 187–88.

8. www.rationalmagic.com/Bursting/Glossary.html (accessed on Feb. 21, 2008).

9. Ibid.

10. As Sheskin suggests, a number of musicals, particularly those "written in the epic-theatre (Brechtian) story telling tradition," have their eleven-o'clock numbers earlier in the play. For instance, Sondheim's *Sweeney Todd* has an "'epiphany' at the end of act 1, which fits the definition perfectly except that it occurs well before the conventional placement near the end of act 2." One can see "how the main character comes to a revelation ('They all deserve to die') and how emotional it is (swinging back and forth from anger at the world to sadness over his loss)" (email communication, Oct. 18, 2007).

11. Sheskin, email communication, Feb. 22, 2008.

12. See Citron, *Opera on Screen,* 56, 102, 227; Newcomb, "Once More," 234; Abbate, *Unsung Voices,* 24, 26, 69, 119–23, 157; Hutcheon and Hutcheon, "Narrativizing the End," 442; and Taruskin, "She Do the Ring," 196.

13. I am deeply indebted to Kang-i Sun Chang for my discussion of the Chinese opera. As she puts it, "the readers (or audiences) always believe that *qu* songs express the genuine feelings of the authors/characters" (email discussion, Oct. 17, 2007).

14. Compare to Yu's observation that "the one continuous thread of argument running through what might be called the Confucian view of the arts is this contradictory desire for spontaneity and calculation, freedom and control" (81–82).

15. Wang, *The Story of the Western Wing,* 120–21, 120nn23–24.

16. Boulton, *The Song in the Story,* 19, 20.

17. Ibid., 20, 24. Note that Boulton's argument is somewhat similar to that of Cleanth Brooks and Robert Penn Warren in *Understanding Poetry* (1938), where they suggest that a Japanese haiku cannot be viewed as providing a "sudden and fleeting insight" into the mind of the speaker. As they see it, to a Western reader using a translation, a certain image may look as a "sudden and fleeting insight" that provides "a profound revelation," but "as the authorities on such poetry tell us, to the mind that is saturated with the rich symbolism of the East, the images of such poems are rich in specific association" (70).

18. Kang-i Sun Chang, *The Evolution of Chinese Tz'u Poetry,* 19. Compare this to Susan Lanser's observation that in lyric poetry we tend to identify the "I" of the narrator with the poem's author ("The 'I' of the Beholder," 207).

19. I am grateful to Evelyn Birge Vitz for reminding me how unreadable the minds in some versions of "Raggle Taggle Gypsy-o" can be.

20. Freydkin, "To Wood."

21. Mead, "The Actress," 54.

22. Sheskin, personal conversation, Oct. 3, 2007. Compare to T. S. Eliot's discussion of the moments when "we touch the border of [the] feelings which only music can express" (Eliot, *On Poetry and Poets,* 87; quoted in Yu, 97).

Nine: Painting Feelings

1. As Fried describes it, "the transparent, slightly distended globe at the tip of his blowpipe seems almost to swell and tremble before our eyes" (*Absorption and Theatricality,* 51).

2. Ibid., 51.

3. Ibid., 157.

4. See the section on problem pictures in chap. 10.

5. Bloom, *How Pleasure Works,* 143–44.

6. Ibid., 139.

7. Schjeldahl, "For Laughs," 84.

8. See Bloom's *How Pleasure Works* for a compelling attempt to bridge that gap.

9. In fact, there have been some interesting speculations on what mirror neurons—a possible neural foundation of our intentionality attribution—might be doing while we observe people in paintings. See, e.g., Freedberg, "Empathy, Motion and Emotion"; and Lindenberger, "Arts in the Brain." For, it could be argued (not an uncontroversial argument, but one worth considering) that we cannot grasp the meaning of any painted gesture unless these neurons are activated. This means that the unconscious mental processes involved in mind reading are constantly interacting with our conscious observations and interpretations. They reinforce and influence each other as we are looking at the painting and thinking about it.

10. For a discussion of surrealism and theory of mind see Zunshine, *Strange Concepts.*

11. Something very similar happens when we read a work of literature that challenges our theory of mind. Imagine a fictional story that opens with two long paragraphs describing a mechanical gadget, a description strikingly lacking in any language of intentionality. If you don't stop reading after the first paragraph and abandon the story for good, you will start asking yourself what kind of narrator we are dealing with here. What are her intentions? What is she trying to achieve with this off-putting opening? Or—a closely related possibility—what kind of author is this? Is she known for writing experimental fiction? Is that what she is up to here—playing with your expectations, seeing how long you will stick it out? And—now the attribution of intentionality shifts to you as the reader—should you trust this author and continue reading? Should you hope that she knows what she is doing and will reward your perseverance with some lovely literary gambit?

In other words, if we can't attribute minds to somebody or something within a narrative known as a work of fiction we go right on attributing them to some entity around the edges of the narrative (i.e., the narrator) or outside the narrative (i.e., the author or ourselves). If there seem to be no mental states in the opening of the story, our mind-reading adaptations may try building on that dreary opening to generate some anyway.

12. Reza, "Art," 4.

13. Ibid., 16–17.

14. Once more, mine is a deliberately limited approach to abstract art, focusing only on its sociocognitive aspects. For an insightful and very different perspective on cognition and the aesthetics of abstract art see Ramachandran, *The Tell-Tale Brain*, esp. chaps. 7 and 8.

15. In the context of the argument to follow, this raises an interesting question: is it ever possible for a character to look directly at the viewer and still be perceived as transparent? It seems not, but I will be happy to stand corrected if you find any examples to the contrary.

16. See Chauvet et al., *Chauvet Cave*. Also, for an insightful argument on "making special," see Dissanayake, *What Is Art For?*

17. Literary criticism does it, too, of course, making us imagine, for example, what Freud, or Foucault, or Jameson would think of this or that text. For a discussion see Zunshine, "Cognitive Alternatives to Interiority."

18. Fried, *Absorption and Theatricality*, 50. See also Fried's discussion of time in modernist art, which, as he argues, seeks to "defeat theater . . . by virtue of [its] presentness and instantaneousness" ("Art and Objecthood," 167).

19. Or as Fried puts it, amplifying the view of Diderot and his contemporaries, a painting has "first to attract . . . and then to arrest . . . and finally to enthrall . . . the beholder, that is, a painting [has] to call to someone, bring him to a halt in front of itself, and hold him there as if spellbound and unable to move" (*Absorption and Theatricality*, 92).

20. As Fried puts it, "only by establishing the fiction of [the beholder's] absence or nonexistence could his actual placement before and his enthrallment by the paint-

ing be secured" (*Absorption and Theatricality*, 103). Compare this to Fried's argument in his earlier essay, "Art and Objecthood" (1967), in which he called his readers' "attention to the utter pervasiveness—the virtual universality—of the sensibility or mode of being . . . corrupted or perverted by theater" (161). "Art and Objecthood" focuses on works of modernist art that "defeat theater" by their quality of "presentness," that is, by their apparent ability to just *be* there independently of the perspective of the beholder. As Fried puts it in the famous last sentence of that essay, "presentness is grace" (168).

21. And, no, I will not construct a glib "historicist" argument of "influence" (e.g., an argument about the French absorptive paintings somehow "informing" the English novel, or vice versa). Each genre can be historicized based on its own cultural history and perennial mind-reading tensions. Although exploration of cross-genre influences is generally a fruitful endeavor, such influences should not be overrated.

22. Fried, *Absorption and Theatricality*, 61.

23. Ibid., 159.

24. Ibid., 156.

25. Ibid., 152.

26. Ibid., 61.

27. Ibid., 176.

28. Ibid., 13.

29. On fiction and sociocognitive complexity see Zunshine, "What to Expect."

30. Fried, *Absorption and Theatricality*, 55–56.

31. Ibid., 55.

32. Ibid., 55.

33. For a fascinating discussion of blushing "as an honest signal of how one feels" and its treatment in fiction, specifically from an evolutionary perspective, see William Flesch's *Comeuppance*, 103–4.

34. *OED*, 2nd online edition (1989), s.v. "sentimental."

35. Fried, *Absorption and Theatricality*, 55.

36. Incidentally, by "convincing" I do not mean "realistic"; for example, an otherworldly setup of a science fiction story can be completely socially convincing but not realistic in the conventional sense of the word.

37. For a very interesting instance of autobiographical probing of how one's former musical tastes would be considered sentimental today see Terry Castle's memoir, *The Professor*. Castle talks about the songs she liked in the 1970s, which she now sees as "treaclefest[s]," redolent of their "genre's gauzy inanities" (159), even as she admits to herself that they still have the power to hit "the thirty-year-old emotional love-spot with warmth and precision" (165).

38. For a cognitive take on the discussion of Fried's resistance to the theory of influence (which I don't discuss here) see Zunshine, "Theory of Mind and Michael Fried's *Absorption and Theatricality*."

Ten: Painting Mysteries

1. Kern, *Eyes of Love,* 7.

2. I am using the titles of paintings from Kern's lavishly illustrated study.

3. Kern discusses these paintings in some detail; see ibid., 71, 93, and 65, respectively. I have purposely chosen proposal compositions of very different styles.

4. Fletcher, *Narrating Modernity,* 62.

5. Ibid., 63.

6. Ibid., 129.

7. Ibid., 62.

8. It's true that in some societies it is now possible to check paternity by testing DNA, but this invention is too recent to have any influence on the psychology of mating.

Bibliography

||||||||||||||||||||||||||||||||||||||||||||||||||||||||||||||||||||||||||||||||||||||||||||||||||||||||||||||||||||||||||||||||||||||||

Abbate, Carolyn. *Unsung Voices: Opera and Musical Narrative in the Nineteenth Century.* Princeton, NJ: Princeton University Press, 1991.

Abbott, Porter. "Conversion in an Age of Darwinian Gradualism." *Storyworlds* 2 (2010): 1–18.

———. *The Fine Art of Failure: Narrative and the Unknowable.* In progress.

———. "Reading Intended Meaning Where None Is Intended." *Poetics Today* 32 no. 3 (2011): 459–85.

Aldama, Frederick Luis. "Race, Cognition, and Emotion: Shakespeare on Film." *College Literature* 33, no. 1 (winter 2006): 197–213.

———, ed. *Toward a Cognitive Theory of Narrative Acts.* Austin: University of Texas Press, 2010.

Anderson, Joseph D., and Barbara Fisher Anderson. *Moving Image Theory: Ecological Considerations.* Carbondale: Southern Illinois University Press, 2007.

Arendt, Hannah. *The Life of the Mind.* New York: Harcourt Brace Jovanovich, 1978.

Auerbach, Erich. *Mimesis.* Princeton, NJ: Princeton University Press, 1991.

Austen, Jane. *Emma.* New York: Bantam, 1981.

———. *Persuasion.* Ed. Gillian Beer. New York: Penguin, 2003.

———. *Pride and Prejudice.* New York: Bantam, 2003.

Austin, Michael. *Useful Fictions: Evolution, Anxiety, and the Origins of Literature.* Lincoln: University of Nebraska Press, 2011.

Baillargeon, R., Z. He, P. Setoh, R. Scott, and D. Yang. "The Development of False-Belief Understanding and Why It Matters." In *The Development of Social Cognition,* ed. M. Banaji and S. Gelman. Oxford: Oxford University Press, forthcoming.

Bargh, John A., and Lawrence E. Williams. "On the Nonconscious Regulation of Emotion." In *Handbook of Emotion Regulation,* ed. James Gross, 429–45. New York: Guilford, 2007.

Barkow, Jerome H., Leda Cosmides, and John Tooby. eds. *The Adapted Mind: Evolutionary Psychology and the Generation of Culture.* New York: Oxford University Press, 1992.

Barnes, Jennifer. "Fiction and Empathy: Narrative Cognition in Autism." A talk in the series "Current Work in Developmental Psychology." Department of Psychology, Yale University, Jan. 14, 2009.

Barnes, Jennifer, R. Li, Simon Baron-Cohen, and Paul Bloom. "Reading Preferences, Empathy, and Autism." In preparation.

Baron-Cohen, Simon. *Mindblindness: An Essay on Autism and Theory of Mind.* Cambridge: MIT Press, 1995.

Barrett, Lisa Feldman, Kevin N. Ochsner, and James J. Gross. "On the Automaticity of Emotion." In *Social Psychology and the Unconscious: The Automaticity of Higher Mental Processes,* ed. John A. Bargh, 173–218. New York: Psychology Press of Taylor and Francis Group, 2007.

Bauman, Melissa D., Eliza Bliss-Moreau, Christopher J. Machado, and David G. Amaral. "The Neurobiology of Primate Social Behavior." In Decety and Cacioppo, *The Oxford Handbook of Social Neuroscience,* 683–701.

Belin, Pascal. "'Hearing Voices': Neurocognition of the Human Voice." In Decety and Cacioppo, *The Oxford Handbook of Social Neuroscience,* 378–393.

Bering, Jesse M. *The Belief Instinct: The Psychology of Souls, Destiny, and the Meaning of Life.* New York: Norton, 2011.

———. "The Existential Theory of Mind." *Review of General Psychology* 6, no. 1 (2002): 3–24.

Blair, Robert James Richard. "Theory of Mind, Autism, and Emotional Intelligence." In *The Wisdom in Feeling: Psychological Processes in Emotional Intelligence,* ed. Lisa Feldman Barrett and Peter Salovey, 406–34. New York and London: The Guilford Press, 2002.

Bloom, Paul. *Descartes' Baby: How the Science of Child Development Explains What Makes Us Human.* New York: Basic Books, 2004.

———. *How Pleasure Works. The New Science of Why We Like What We Like.* New York: Norton, 2010.

Booth, Wayne C. *The Rhetoric of Fiction.* Chicago: University of Chicago Press, 1961.

Borenstein, Elhanan, and Eytan Ruppin. "The Evolution of Imitation and Mirror Neurons in Adaptive Agents." *Cognitive Systems Research* 6, no. 3 (2005): 229–42.

Bortolussi, Marisa, and Peter Dixon. *Psychonarratology: Foundations for the Empirical Study of Literary Response.* Cambridge, UK: Cambridge University Press, 2003.

Boulton, Maureen Barry McCann. *The Song in the Story: Lyric Insertions in French Narrative Fiction, 1200–1400.* Philadelphia: University of Pennsylvania Press, 1993.

Branigan, Edward. *Projecting a Camera: Language-Games in Film Theory.* New York: Routledge, 2006.

Brooks, Cleanth, and Robert Penn Warren. *Understanding Poetry.* 4th ed. Fort Worth, TX: Harcourt Brace College Publishers, 1976.

Burney, Frances. *Evelina; or, The History of a Young Lady's Entrance into the World (1778).* New York: Modern Library, 2001.

Butler, Emily A., and James J. Gross. "Hiding Feelings in Social Contexts: Out of Sight Is Not Out of Mind." In *The Regulation of Emotion,* ed. Pierre Philippot and

Robert S. Feldman, 101–26. Mahwah, NJ: Erlbaum, 2004.

Butler, Emily A., T. L. Lee, and James J. Gross. "Emotion Regulation and Culture: Are the Social Consequences of Emotion Suppression Culture-Specific?" *Emotion* 7 (2007): 30–48.

Butte, George. *I Know That You Know That I Know: Narrating Subjects from "Moll Flanders" to "Marnie."* Columbus: Ohio State University Press, 2004.

Byrne, Richard W., and Andrew Whiten. "The Emergence of Metarepresentation in Human Ontogeny and Primate Phylogeny." In Whiten, *Natural Theories of Mind*, 267–82.

———. *Machiavellian Intelligence: Social Expertise and the Evolution of Intellect in Monkeys, Apes, and Humans.* New York: Oxford University Press, 1988.

Carroll, Noël. *The Philosophy of Motion Pictures.* Malden, MA: Blackwell, 2008.

Carruthers, Peter. *The Opacity of Mind: An Integrative Theory of Self-Knowledge.* New York: Oxford University Press, 2011.

Caserio, Robert L., and Clement Hawes, eds. *The Cambridge History of the English Novel.* New York: Cambridge University Press, 2012.

Castle, Terry. *The Professor and Other Writings.* New York: Harper Collins, 2010.

Cavell, Stanley. *Pursuits of Happiness: The Hollywood Comedy of Remarriage.* Cambridge, MA: Harvard University Press, 1981.

Chang, Kang-i Sun. *The Evolution of Chinese Tz'u Poetry: From Late T'ang to Northern Sung.* Princeton, NJ: Princeton University Press, 1980.

Chauvet, Jean-Marie, et al. *Chauvet Cave: The Discovery of the World's Oldest Paintings.* London: Thames and Hudson, 1996.

Citron, Marcia J. *Opera on Screen.* New Haven, CT: Yale University Press, 2000.

Cohn, Dorrit. *Transparent Minds: Narrative Modes for Presenting Consciousness in Fiction.* Princeton, NJ: Princeton University Press, 1978.

Crane, Mary Thomas. "Surface, Depth, and the Spatial Imaginary. A Cognitive Reading of the Political Unconscious." *Representations* 108 no. 1 (2009): 76–97.

Csibra, G. "Goal Attribution to Inanimate Agents by 6.5-Month-Old Infants." *Cognition* 107 (2008): 705–17.

Currie, Gregory. *Image and Mind: Film, Philosophy and Cognitive Science.* New York: Cambridge University Press, 2008.

Decety, Jean, and John T. Cacioppo, eds. *The Oxford Handbook of Social Neuroscience.* New York: Oxford University Press, 2011.

Defoe, Daniel. *The Life and Strange Surprizing Adventures of Robinson Crusoe, or York, Mariner . . .* Ed. with an introduction and notes by J. Donald Crowley. New York: Oxford University Press, 1998.

Denby, David. "Soldiers." *New Yorker,* Sept. 24, 2007, 188–89.

Dennett, Daniel C. *Consciousness Explained.* Boston: Little, Brown, 1991.

———. *The Intentional Stance.* Cambridge, MA: MIT Press, 1989.

Diderot, Denis. *The Paradox of Acting.* Trans. with annotations from Diderot's "Paradoxe sur le comédien» by Walter Herries Pollock. London: Chatto & Windus, 1883.

Dissanayake, Ellen. *What Is Art For?* Seattle: University of Washington Press, 1988.

Dowd, Maureen. "Cool Hand Barack." *New York Times,* May 3, 2011.

Dunbar, Robin. "Evolutionary Basis of the Social Brain." In Decety and Cacioppo, *The Oxford Handbook of Social Neuroscience,* 28–38.

Dunne, Michael. *American Film: Musical Themes and Forms.* Jefferson, NC: McFarland, 2004.

Dutton, Denis. *The Art Instinct: Beauty, Pleasure, and Human Evolution.* New York: Bloomsbury, 2010.

Easterlin, Nancy. *A Biocultural Approach to Literary Theory and Interpretation.* Baltimore: Johns Hopkins University Press, 2012.

Ekman, Paul. "Strong Evidence for Universals in Facial Expressions: A Reply to Russell's Mistaken Critique." *Psychological Bulletin* 115 (1994): 268–87.

Ekman, Paul, and Alan J. Fridlund. "Assessment of Facial Behavior in Affective Disorders." *Depression and Expressive Behavior,* ed. J. D. Maser, 37–56. Hillsdale, NH: Erlbaum, 1985.

Eliot, T. S. *On Poetry and Poets.* New York: Noonday Press, 1961.

Ellis, Jack C., and Betsy A. McLane. *A New History of Documentary Film.* New York: Continuum, 2005.

Feuer, Jane. *The Hollywood Musical.* Bloomington: Indiana University Press, 1982.

Fielding, Helen. *Bridget Jones: The Edge of Reason.* New York: Penguin, 1999.

Fielding, Henry. *Tom Jones.* 1749. Ed. John Bender and Simon Stern. Oxford: Oxford University Press, 1996.

Fielding, Sarah. *The History of Ophelia.* London: R. Baldwin, 1760.

Finocchiaro, Peter. "Obama's Poker Face." *Salon.com,* May 2, 2011.

Flesch, William. *Comeuppance: Costly Signaling, Altruistic Punishment, and Other Biological Components of Fiction.* Cambridge, MA: Harvard University Press, 2007.

Fletcher, Pamela M. *Narrating Modernity: The British Problem Picture, 1895–1914.* Aldershot, UK: Ashgate, 2003.

Fludernik, Monika. "1050–1500: Through a Glass Darkly; or, The Emergence of Mind in Medieval Literature." In Herman, *The Emergence of Mind,* 69–100.

Freedberg, David. "Empathy, Motion and Emotion." In *Wie sich Gefühle Ausdruck verschaffen: Emotionen in Nahsicht,* ed. K. Herding and A. Krause Wahl, 17–51. Berlin: Driesen, 2007.

Freydkin, Donna. "To Wood, Fame Is 'High School.'" *USA Today,* Sept. 24, 2007.

Fridlund, Alan J. "Evolution and Facial Action in Reflex, Social Motive, and Paralanguage." *Biological Psychology* 32 (1991): 3–100.

Fried, Michael. *Absorption and Theatricality: Painting and Beholder in the Age of Diderot.* Berkeley: University of California Press, 1980.

———. "Art and Objecthood." 1967. Repr. in *Art and Objecthood: Essays and Reviews,* 148–72. Chicago: University of Chicago Press, 1998.

———. "Barthes's Punctum." *Critical Inquiry* 31 (spring 2005): 539–74.

Frith, Christopher D., and Uta Frith. "Social Cognition in Humans." *Current Biology,* no. 17 (2007): R724–32.

Gallagher, Catherine, and Stephen Greenblatt. *Practicing New Historicism.* Chicago: University of Chicago Press, 2000.

Gervais, Ricky, and Stephen Merchant. *The Office: The Complete First Series.* 2 disks. BBC Worldwide Americas and Warner Home Video, 2003.

Gevorkian, Natalia. "Ne povod dlia tanzev." *Gazeta.ru,* May 5, 2011, www.gazeta.ru/column/gevorkyan/3604309.shtml (accessed May 27, 2011).

Goffman, Erving. *Strategic Interaction.* New York: Ballantine, 1969.

Goldman, Alvin I. *Simulating Minds: The Philosophy, Psychology, and Neuroscience of Mindreading.* New York: Oxford University Press, 2006.

Gombrich, E. H. *Art and Illusion: A Study in the Psychology of Pictorial Representation.* 3rd ed. Princeton, NJ: Princeton University Press, 1969.

Gomez, Juan C. "Visual Behavior as a Window for Reading the Mind of Others in Primates." In Whiten, *Natural Theories of Mind,* 195–208.

Gross, James J., and Dacher Keltner, eds. *Cognition and Emotion* 13, no. 5 (1999).

Grossman, Tobias, Tricia Striano, and Angela D. Friederici. "Developmental Changes in Infants' Processing of Happy and Angry Facial Expressions: A Neurobehavioral Study." *Brain and Cognition* 64, no. 1 (2007): 30–41.

Guthrie, Stewart Elliott. *Faces in the Clouds: A New Theory of Religion.* New York: Oxford University Press, 1995.

Hans, Jonas. "The Nobility of Sight: A Study in the Phenomenology of Senses." *Philosophy and Phenomenological Research* 14, no. 5 (1954): 507–19.

Hart, F. Elizabeth. "The Epistemology of Cognitive Literary Studies." *Philosophy and Literature* 25, no. 2 (2001): 314–34.

Hegel, Georg Wilhelm Friedrich. *The Phenomenology of Mind.* Trans. with an introduction and notes by J. B. Baillie. 2nd ed. London: George Allen and Unwin, 1977.

Hellman, Robert, and Richard O'Gorman. *Fabliaux: Ribald Tales from the Old French.* New York: Thomas Y. Crowell, 1966.

Hemingway, Ernest. *A Farewell to Arms.* New York: Charles Scribner's Sons, 1929.

Herman, David, ed. *The Emergence of Mind: Representations of Consciousness in Narrative Discourse in English.* Lincoln: University of Nebraska Press, 2011.

———. "Genette Meets Vygotsky: Narrative Embedding and Distributed Intelligence." *Language and Literature* 15, no. 4 (2006): 357–80.

———, ed. *Narrative Theory and the Cognitive Sciences.* Stanford, CA: CSLI, 2003.

———. *Story Logic.* Lincoln: University of Nebraska Press, 2002.

Hickok, Gregory. "Eight Problems for the Mirror Neuron Theory of Action Understanding in Monkeys and Humans." *Journal of Cognitive Neuroscience* 21, no. 7 (2008): 1229–43

Hogan, Patrick Colm. *Cognitive Science, Literature, and the Arts: A Guide for Humanists.* New York: Routledge, 2003.

———. *Empire and Poetic Voice: Cognitive and Cultural Studies.* Albany: SUNY Press, 2004.

———. "Literary Universals." *Poetics Today* 18, no. 2 (1997): 223–49.

———. *The Mind and Its Stories: Narrative Universals and Human Emotion.* Cambridge, UK: Cambridge University Press, 2003.

———. *Understanding Nationalism: On Narrative, Neuroscience, and Identity.* Columbus: Ohio State University Press, 2009.

Hornby, Nick. *Fever Pitch.* New York: Penguin, 1992.

———. *How to Be Good.* New York: Riverhead, 2001.

Hutcheon, Linda, and Michael Hutcheon. "Narrativizing the End: Death and Opera." In *A Companion to Narrative Theory,* ed. James Phelan and Peter J. Rabinowitz, 441–50. Malden, MA: Blackwell, 2005.

Jackson, Tony. "Issues and Problems in the Blending of Cognitive Science, Evolutionary Psychology, and Literary Study." *Poetics Today* 23, no. 1 (2002): 161–79.

Johnson, Susan, Virginia Slaughter, and Susan Carey. "Whose Gaze Would Infants Follow? The Elicitation of Gaze Following in 12-Month-Olds." *Developmental Science* 1 (1998): 233–38.

Jones, Wendy S. "*Emma,* Gender, and the Mind-Brain." *ELH* 75, no. 2 (2008): 315–43.

Kaufmann, Walter. *Tragedy and Philosophy.* New York: Doubleday, 1968.

Keen, Suzanne. *Empathy and the Novel.* Oxford University Press, 2007.

———. "Strategic Empathizing: Techniques of Bounded, Ambassadorial, and Broadcast Narrative Empathy." *Deutsche Vierteljahrs Schrift* 82, no. 3 (Sept. 2008): 477–93.

———. *Thomas Hardy's Brains.* In progress.

Keenan, Julian Paul, Hanna Oh, and Franco Amati. "An Overview of Self-Awareness and the Brain." In Decety and Cacioppo, *The Oxford Handbook of Social Neuroscience,* 314–24.

Kelly, David J., et al. "Social Experience Does Not Abolish Cultural Diversity in Eye Movements." *Frontiers in Cultural Psychology* 2, no. 95. Published online May 18, 2011, www.frontiersin.org/cultural_psychology/10.3389/fpsyg.2011.00095/full (accessed Dec. 2, 2011).

Keough, Peter. "Shades of Grey: Tending Again to the Maysleses' Gardens." http://weeklywire.com/ww/09-08-98/boston_movies_1.html (accessed August 18, 2007).

Kern, Stephen. *Eyes of Love: The Gaze in English and French Paintings and Novels, 1840–1900.* London: Reaktion, 1996.

Keysers, Christian, Marc Thioux, and Valeria Gazzola. "The Mirror Neuron System and Social Cognition." In Decety and Cacioppo, *The Oxford Handbook of Social Neuroscience,* 525–541.

Kleiner, Fred S. *Gardner's Art through the Ages: A Global History, Enhanced Thirteenth Edition,* Vol. 1. Boston: Wadsworth, 2010.

Kramnick, Jonathan. "Some Thoughts on Print Culture and the Emotions." *The Eighteenth Century: Theory and Interpretation* 50 no. 2–3 (2009): 263–67.

Kurzban, Robert. *Why Everyone (Else) Is a Hypocrite: Evolution and the Modular Mind.* Princeton, NJ: Princeton University Press, 2010.

Lahr, John. "Solos and Solitaries." *New Yorker,* March 21, 2005, 88–89.

Lane, Anthony. "Miles to Go." *New Yorker,* Jan. 31, 2011, 82–83.

Lanser, Susan. "The 'I' of the Beholder: Equivocal Attachments and the Limits of Structuralist Narratology." In *A Companion to Narrative Theory,* ed. James Phelan and Peter J. Rabinowitz, 206–19. Malden, MA: Blackwell, 2005.

Levin, David Michael. "Introduction." In *Modernity and the Hegemony of Vision,* ed. David Michael Levin, 1–29. Berkeley: University of California Press, 1993.

Lindenberger, Herbert. "Arts in the Brain; or, What Might Neuroscience Tell Us?" In *Toward a Cognitive Theory of Literary Acts,* ed. Frederick Aldama. Austin: University of Texas Press, 2010. 13–35.

Longin, Sheryl. "Confessions of a Reality Junkie." *Pajamas Media,* August 11, 2007, http://pajamasmedia.com/blog/confessions_of_a_reality_junki (accessed July 13, 2008).

Lopate, Phillip. "Portrait of My Body." In *Getting Personal: Selected Writings,* 327–34. New York: Basic Books, 2003.

Luo, Y., and R. Baillargeon. "Can a Self-Propelled Box Have a Goal? Psychological Reasoning in 5-Month-Old Infants." *Psychological Science* 16 (2005): 601–8.

Mamet, David. *On Directing Film.* New York: Viking, 1991.

Mazzetti, Mark, Helene Cooper, and Peter Baker. "Behind the Hunt for Bin Laden." *New York Times,* May 2, 2001.

McClure, Erin B. "A Meta-analytic Review of Sex Differences in Facial Expression Processing and Their Development in Infants, Children, and Adolescents." *Psychological Bulletin* 126 (2000): 424–53.

McConachie, Bruce. *American Theater in the Culture of the Cold War: Producing and Contesting Containment, 1947–1962.* Iowa City: University of Iowa Press, 2003.

———. *Engaging Audiences: A Cognitive Approach to Spectating in the Theatre.* New York: Palgrave Macmillan, 2008.

McGinn, Colin. *The Power of Movies: How Screen and Mind Interact.* New York: Pantheon, 2005.

McMillin, Scott. *The Musical as Drama: A Study of the Principles and Conventions behind Musical Shows from Kern to Sondheim.* Princeton, NJ: Princeton University Press, 2006.

Mead, Rebecca. "The Actress." *New Yorker,* March 2, 2009, 52–59.

Merleau-Ponty, Maurice. *Phenomenology of Perception.* Trans. Colin Smith. London: Routledge and Kegan Paul, 1962.

Michaels, Walter Benn. *The Shape of the Signifier: 1967 to the End of History.* Princeton, NJ: Princeton University Press, 2004.

Miller, Scott. *From "Assassins" to "West Side Story": The Director's Guide to Musical Theatre.* Portsmouth, NH: Heinemann, 1996.

Moretti, Franco. "The Slaughterhouse of Literature." *Modern Language Quarterly* 61, no. 1 (March 2000): 207–28.

Mullan, John. *Sentiment and Sociability: The Language of Feeling in the Eighteenth Century.* Oxford: Clarendon, 1988.

Nettle, Daniel. "Emphasizing and Systemizing: What Are They, and What They Contribute to Our Understanding of Psychological Sex Differences." *British Journal
of Psychology* 98, no. 2 (2007): 237–55.

———. "Psychological Profiles of Professional Actors." *Personality and Individual
Differences* 40 (2006): 375–83.

Newcomb, Anthony. "Once More 'Between Absolute and Program Music': Schu-
mann's Second Symphony." *19th-Century Music* 7 (1984): 233–50.

Nussbaum, Martha C. *Upheavals of Thought: The Intelligence of Emotions.* Cambridge, UK: Cambridge University Press, 2001.

Palahniuk, Chuck. *Fight Club.* New York: Henry Holt, 1996.

Palmer, Alan. *Fictional Minds.* Lincoln: University of Nebraska Press, 2004.

———. *Social Minds in the Novel.* Columbus: Ohio State University Press, 2010.

———. "Storyworlds and Groups." In Zunshine, *Introduction to Cognitive Cultural
Studies,* 178–92.

"The Past, Present, and Future of Reality Television." Museum of Television & Radio
Seminar Series. Sept. 25, 2003 (PAC). Moderated by Barbara Dixon. Panelists:
Mike Darnell, Mike Fleiss, Ghen Maynard, Jonathan Murray, Arnold Shapiro,
Scott A. Stone, and Andrea Wong.

Pemberton, Gayle. *The Hottest Water in Chicago: Notes of a Native Daughter.* Ha-
nover and London: Wesleyan University Press, 1992.

Pentland, Alex. *Honest Signals: How They Shape Our World.* Cambridge, MA: MIT
Press, 2008.

Persson, Per. *Understanding Cinema: A Psychological Theory of Moving Imagery.*
Cambridge, UK: Cambridge University Press, 2003.

Phelan, James. *Living to Tell about It: A Rhetoric and Ethics of Character Narration.*
Ithaca, NY: Cornell University Press, 2004.

Phelan, Peggy. "Reciting the Citation of Others; or, A Second Introduction." In *Acting Out: Feminist Performances,* ed. Lynda Hart and Peggy Phelan, 13–31. Ann
Arbor: University of Michigan Press, 1993.

Pierpont, Claudia Roth. "The Player Kings." *New Yorker,* Nov. 19, 2007, 70–79.

Plantinga, Carl, and Greg M. Smith. *Passionate Views: Film, Cognition, and Emotion.*
Baltimore: Johns Hopkins University Press, 1999.

Premack, David, and Verena Dasser. "Perceptual Origins and Conceptual Evidence
for Theory of Mind in Apes and Children." In Whiten, *Natural Theories of
Mind,* 253–66.

Priborkin, Klarina. "Cross-Cultural Mind Reading: Challenging the Universality of
the Unspoken in Maxine Hong Kingston's *The Woman Warrior.*" Paper presented at Literature and Cognitive Science Conference. University of Connecti-
cut, Storrs, April 8, 2006.

Quigley, Rachel. "Maybe I Just Coughed: Hillary Clinton Downplays Expression of
Shock in Situation Room Photo Claiming It Was 'Spring Allergies.'" *Daily Mail
Online,* May 5, 2011.

Ramachandran, V. S. *A Brief Tour of Human Consciousness: From Impostor Poodles to Purple Numbers.* New York: Pi Press, 2004.
———. *The Tell-Tale Brain.* New York: Norton, 2011.
Remnick, David. "Exit Bin Laden." *New Yorker,* May 16, 2011, 35–36.
Reza, Yasmina. *"Art."* Trans. Christopher Hampton. New York: Faber and Faber, 1997.
Richardson, Alan. *British Romanticism and the Science of the Mind.* Cambridge, UK: Cambridge University Press, 2001.
———. "Facial Expression Theory from Romanticism to the Present." In Zunshine, *Introduction to Cognitive Cultural Studies,* 65–83.
———. *The Neural Sublime: Cognitive Theories and Romantic Texts.* Baltimore: Johns Hopkins University Press, 2010.
———. "Studies in Literature and Cognition: A Field Map." In Richardson and Spolsky, *The Work of Fiction,* 1–30.
Richardson, Alan, and Ellen Spolsky, eds. *The Work of Fiction: Cognition, Culture, and Complexity.* Aldershot, UK: Ashgate, 2004.
Richardson, Samuel. *Clarissa, or, The History of a Young Lady.* Ed. Angus Ross. New York: Penguin, 1985.
Rizzolatti, Giacomo, Leonardo Fogassi, and Vittoriao Gallese. "Neuropsychological Mechanisms Underlying the Understanding and Imitation of Action." *Nature Reviews Neuroscience* 2, no. 9 (2001): 661–70.
Roach, Joseph. "Culture and Performance in the Circum-Atlantic World." In *Performativity and Performance,* ed. Andrew Parker and Eve Kosofsky Sedgwick, 45–63. New York: Routledge, 1995.
Rodgers, Richard, and Oscar Hammerstein II. *South Pacific.* 1958. Twentieth Century Fox Home Entertainment, 2006.
Rousseau, Jean-Jacques. *Emile: or On Education.* Translated by Allan Bloom. New York: Basic Books, 1979.
Russell, James A., Jo-Anee Bachorowski, and Jose-Miguel Fernandez-Dols. "Facial and Vocal Expressions of Emotion." *Annual Review of Psychology* 54 (2003): 329–49.
Savarese, Emily Thornton, and Ralph James Savarese, eds. *Autism and the Concept of Neurodiversity.* Special issue of *Disability Studies Quarterly* 30, no. 1 (2010).
Saxe, Rebecca. "Why and How to Study Theory of Mind with fMRI." *Brain Research* 1079 (2006): 57–65.
Saxe, Rebecca, and Nancy Kanwisher. "People Thinking about Thinking People: The Role of the Temporo-parietal Junction in 'Theory of Mind.'" *Neuroimage* 19 (2003): 1835–42.
Scarry, Elaine. *The Body in Pain: The Making and Unmaking of the World.* New York: Oxford University Press, 1985.
———. *Dreaming by the Book.* New York: Farrar, Straus, Giroux, 1999.
Schjeldahl, Peter. "For Laughs." *New Yorker,* May 23, 2011, 84–85.

Schultz, Robert T. "Developmental Deficits in Social Perception in Autism: The Role of the Amygdala and Fusiform Face Area." *International Journal of Developmental Neuroscience* 23 (2005): 125–41.

Seyfarth, Robert M., and Dorothy L. Cheney. "Signalers and Receivers in Animal Communication." *Annual Review of Psychology* 54 (2003): 145–73.

Shakespeare, William. *Cymbeline*. Ed. Roger Warren. Oxford: Oxford University Press, 1998.

Shany-Ur, Tal, and Simone G. Shamay-Tsoory. "Theory of Mind Deficits in Neurological Patients." In Decety and Cacioppo, *The Oxford Handbook of Social Neuroscience*, 935–45.

Siddons, Henry. *Practical Illustrations of Rhetorical Gesture and Action, Adapted to the English Drama* ... 1807. London: Sherwood, Neely, and Jones, 1822.

Singer, Tania, Daniel Wolpert, and Chris Frith. "Introduction: The Study of Social Interactions." In *The Neuroscience of Social Interaction: Decoding, Imitating, and Influencing the Actions of Others*, ed. Christopher D. Frith and Daniel Wolpert, xiii–xxvii. Oxford: Oxford University Press, 2004.

Song, H., and R. Baillargeon. "Infants' Reasoning about Others' False Perceptions." *Developmental Psychology* 44 (2008): 1789–95.

Song, H., K. Onishi, R. Baillargeon, and C. Fisher. "Can an Actor's False Belief Be Corrected by an Appropriate Communication? Psychological Reasoning in 18.5-Month-Old Infants." *Cognition* 109 (2008): 295–315.

Sorce, J. F., R. N. Emde, J. Campos, and M. D. Klinnert. "Maternal Emotional Signaling: Its Effects on the Visual Cliff Behavior of 1-Year-Olds." *Developmental Psychology* 21 (1985): 195–200.

Sperber, Dan. *Explaining Culture: A Naturalistic Approach*. Oxford: Blackwell, 1997.

Spolsky, Ellen. "Cognitive Literary Historicism: A Response to Adler and Gross." *Poetics Today* 24, no. 2 (2003): 161–83.

———. "Darwin and Derrida: Cognitive Literary Theory as a Species of Poststructuralism." *Poetics Today* 23, no. 1 (2002): 43–62.

———. "Elaborated Knowledge: Reading Kinesis in Pictures." *Poetics Today* 17, no. 2 (1996): 157–80.

———. *Gaps in Nature: Literary Interpretation and the Modular Mind*. Albany: State University of New York Press, 1993.

———. "Narrative as Nourishment." In Aldama, *Toward a Theory of Narrative Acts*, 37–60.

———. "Purposes Mistook: Failures Are More Tellable." Talk delivered at the panel on "Cognitive Approaches to Narrative" at the annual meeting of the Society for the Study of Narrative, Burlington, VT, 2004.

———. *Satisfying Skepticism: Embodied Knowledge in the Early Modern World*. Aldershot, UK: Ashgate, 2001.

———. *Word vs Image: Cognitive Hunger in Shakespeare's England*. Basingstoke: Palgrave Macmillan, 2007.

Stafford, Barbara Maria. *Echo Objects: The Cognitive Work of Images.* Chicago: University of Chicago Press, 2007.

———. "Romantic Systematics and the Genealogy of Thought. The Formal Roots of a Cognitive History of Images." *Configurations* 12, no. 3 (2004): 315–48.

Starr, G. Gabrielle. *Feeling Beauty: Aesthetic Perception in the Brain and in the Arts.* Under consideration.

———. "Multisensory Imagery." In Zunshine, *Introduction to Cognitive Cultural Studies,* 275–91.

———. "Poetic Subjects and Grecian Urns: Close Reading and the Tools of Cognitive Science." *Modern Philology* 105, no. 1 (2007): 48–61.

Sternberg, Meir. *Expositional Modes and Temporal Ordering in Fiction.* Baltimore: Johns Hopkins University Press, 1978.

Stiller, James, and Robin I. M. Dunbar. "Perspective-Taking and Social Network Size in Humans." *Social Networks* 29 (2007): 93–104.

Stone, Valerie E., and Catherine A. Hynes. "Real-World Consequences of Social Deficits: Executive Functions, Social Competencies, and Theory of Mind in Patients with Ventral Frontal Damage and Traumatic Brain Injury." In Decety and Cacioppo, *The Oxford Handbook of Social Neuroscience,* 455–76.

Surian, Luca, Stefania Caldi, and Dan Sperber. "Attribution of Beliefs by 13-Month-Old Infants." *Psychological Science* 18, no. 7 (2007): 580–86.

Talbot, Margaret. "Duped." *New Yorker,* July 2, 2007, 52–61.

Tapper, Jake. "Hillary Clinton Explains Famous Osama Bin Laden Raid Photo." *ABC News,* May 5, 2011.

Taruskin, Richard. "She Do the Ring in Different Voices." *Cambridge Opera Journal* 4 (1991): 187–97.

Tiedens, Larissa Z. "Anger and Advancement versus Sadness and Subjugation: The Effect of Negative Emotion Expressions on Social Status Conferral." *Journal of Personality and Social Psychology* 80 (2001): 285–93.

Todorov, Alex, Chis P. Said, Andrew D. Engell, and Nikolaas N. Oosterhof. "Understanding Evaluation of Faces on Social Dimensions." *Trends in Cognitive Sciences* 12, no. 12 (2008): 455–60.

Tolstoy, Leo. *Anna Karenina: A Novel in Eight Parts.* Trans. by Richard Pevear and Larissa Volokhonsky. London: Allen Lane, 2000.

Tolstoy, Lev Nikolaevitch. *Anna Karenina: Roman v vos'mi chastiach.* Chasti pervaia-chetvertaya. Parts 1–4. Tula: Priokskoe Knizhnoie Izdatel'stvo, 1983.

Tooby, John, and Leda Cosmides. "Consider the Source: The Evolution of Adaptations for Decoupling and Metarepresentations." In *Metarepresentations: A Multidisciplinary Perspective,* ed. Dan Sperber, 53–116. New York: Oxford University Press, 2000.

———. "Evolutionary Psychology, Ecological Rationality, and the Unification of the Behavioral Sciences." *Behavioral and Brain Sciences* 30, no. 1 (2007): 42–43.

———. "The Psychological Foundations of Culture." In Barkow, Cosmides, and Tooby, *The Adapted Mind,* 19–136.

Triesch, Jochen, Hector Lasso, and Gedeon O. Deak. "Emergence of Mirror Neurons in a Model of Gaze Following." *Adaptive Behavior* 15 (2007): 149–65.

Turner, Mark. *The Literary Mind: The Origins of Thought and Language.* New York: Oxford University Press, 1998.

Wang Shifu. *The Story of the Western Wing.* Edited and translated by Stephen H. West and Wilt L. Idema. Berkeley: University of California Press, 1995.

Westen, Drew. *The Political Brain: The Role of Emotion in Deciding the Fate of the Nation.* New York: Public Affairs, 2007.

Whiten, Andrew, ed. *Natural Theories of Mind: Evolution, Development, and Simulation of Everyday Mindreading.* Oxford, UK: Basil Blackwell, 1991.

Wintonick, Peter, dir. *Cinéma Vérité: Defining the Moment.* Montreal: National Film Board of Canada, 1999.

Woloch, Alex. *The One vs. the Many: Minor Characters and the Space of the Protagonist in the Novel.* Princeton, NJ: Princeton University Press, 2003.

Vermeule, Blakey. "God Novels." In Richardson and Spolsky, *The Work of Fiction,* 147–66.

———. *The Party of Humanity: Writing Moral Psychology in Eighteenth-Century Britain.* Baltimore: Johns Hopkins University Press, 2000.

———. "Satirical Mind Blindness." *Classical and Modern Literature* 22, no. 2 (2002): 85–101.

———. *Why Do We Care about Literary Characters?* Baltimore: Johns Hopkins University Press, 2010.

Vitz, Evelyn Birge. "Tales with Guts: A 'Rasic' Aesthetic in Medieval French Storytelling." *Drama Review* 52, no. 4 (winter 2008): 145–73.

Yu, Anthony C. *Rereading the Stone: Desire and the Making of Fiction in "Dream of the Red Chamber."* Princeton, NJ: Princeton University Press, 1997.

Zebrowitz, Leslie A. *Reading Faces: Window to the Soul.* Boulder, CO: Westview Press, 1997.

Zebrowitz, Leslie A., and Yi Zhang. "The Origins of First Impressions in Animal and Infant Face Perception." In Decety and Cacioppo, *The Oxford Handbook of Social Neuroscience,* 434–44.

Zehme, Bill. *Lost in the Funhouse: The Life and Mind of Andy Kaufman.* New York: Delacorte, 1999.

Zmuda, Bob, with Matthew Scott Hansen. *Andy Kaufman Revealed!* Boston: Little, Brown, 1999.

Zunshine, Lisa. "Cognitive Alternatives to Interiority." In Caserio and Hawes, *The Cambridge History of the English Novel,* 147–62.

———, ed. *Introduction to Cognitive Cultural Studies.* Baltimore: Johns Hopkins University Press, 2010.

———. "Lying Bodies of the Enlightenment: Theory of Mind and Eighteenth-Century Studies." In Zunshine, *Introduction to Cognitive Cultural Studies,* 115–33.

———. "Mind Plus: Sociocognitive Pleasures of Jane Austen's Novels." *Studies in Literary Imagination* 42, no. 2 (fall 2009): 89–109.

———. "1700–1775: Theory of Mind, Social Hierarchy, and the Emergence of Narrative Subjectivity." In Herman, *The Emergence of Mind,* 161–86.

———. *Strange Concepts and the Stories They Make Possible: Cognition, Culture, Narrative.* Baltimore: Johns Hopkins University Press, 2008.

———. "Theory of Mind and Michael Fried's *Absorption and Theatricality:* Notes toward Cognitive Historicism." In Aldama, *Toward a Cognitive Theory of Narrative Acts,* 179–203.

———. "What Is Cognitive Cultural Studies?" In Zunshine, *Introduction to Cognitive Cultural Studies,* 1–33.

———. "What to Expect When You Pick Up a Graphic Novel." *SubStance.* Special issue on graphic narratives, ed. Jared Gardner and David Herman, issue 124, vol. 40, no.1 (2011): 114–34.

———. *Why We Read Fiction: Theory of Mind and the Novel.* Columbus: Ohio State University Press, 2006.

Index

in Hornby's *Fever Pitch*, 72–75; and horse races, 64–66, 69–72; in *How to Be Good*, 56–58; and humiliation, 119–20; in "The Knight Who Made Cunts Speak," 38–39, 40; in life-threatening circumstances, 29, 32, 45, 111–13, 128; in Lubitsch's *Lady Windermere's Fan*, 66–69; in *Merrily We Roll Along*, 134; in musicals, 125–37; in *My Fair Lady*, 132–33; in *New York Times*, 92–93; in *Notorious*, 69–72; in *The Office*, 103–6, 110–11; in opera, 139, 140; and performance, xii, 31, 58, 59, 60, 73, 76, 101, 106, 108, 121, 135–37, 158; in *Persuasion*, 22–23; in photography, 111–13; in *Practical Illustrations*, 63; in *Pride and Prejudice*, 34–36; in proposal compositions, 169–72, 176; in *Quiz Show*, 93–100; in reality TV, 117–21; rules for portraying, 30–33, 38–39, 51, 63, 74, 106, 129, 130, 147, 155–57, 170, 175, 180; and social complexity, 25–28, 113, 118; in songs, 141–43; in *South Pacific*, 126–30; in stand-up comedy, 113–14; in *The Story of the Western Wing*, 139–40; and strategic obstruction of emotions, 80–81; as subverted, 58, 59, 60, 72, 76, 101, 108–11, 132, 158, 161, 180; in *Sunday in the Park with George*, 135–37; and theater, 56–57, 61–63; in *Tom Jones*, 40–42; and unreliable narration, 50–53; in *What Women Want*, 38, 40

Emile (Rousseau), 47–49, 52

emotions: and Chinese opera, 139, 141; and cinéma vérité, 107–9; display of, 4, 7–8; and fiction, 24; and film, 82–90, 96–100; and horse races, 65, 69–70; 92–93; and photography, 111–13; and reality TV, 117–18, 120–21; in real life vs. cultural representations, 27–28, 32; and rule of contrasts, 106; and sadistic benefactors, 45, 47, 49; and sentimentalism, 162, 164–65; and singing, 143; suppression of, 35–36; and theater, 55–56, 58–59

Ernst, Max, 149

ethics, 32, 36, 39, 42, 45–46, 186n21

Evelina (Burney), 60–61, 62

Extreme Makeover: Home Edition (TV series), 118

Feininger, Lyonel, 150

Fellig, Arthur, 111

Feuer, Jane, 191n2

Fever Pitch (film), 74–76

Fever Pitch (Hornby), 72–75, 165

Fielding, Helen, *Bridget Jones: The Edge of Reason*, 37–38, 40

Fielding, Henry, *Tom Jones*, 40–42

Fielding, Sarah, *The History of Ophelia*, 46–47, 49, 52

Fiennes, Ralph, 93, 97

Fight Club (Palahniuk), 49–53

Finocchiaro, Peter, 92

Firth, Colin, 75, 85

Flesch, William, xii, 188n1, 194n33

Fletcher, Pamela M., 173–76, 195nn4–7

Fludernik, Monika, xii, 188n1

Follies (musical), 126

Foucault, Michel, 186n14, 187n10

Frears, Stephen, *The Queen*, 86–90, 96

Freedberg, David, 192n9

Freydkin, Donna, 192n29

Fridlund, Alan J., 32, 186n18

Fried, Michael: *Absorption and Theatricality*, 145–47, 158–61, 166–67, 177, 190n11, 190n15, 193nn18–20, 194nn22–28, 194nn30–32; and rules of transience and contrasts, 156–58; and sentimentalism, 161–66

Frith, Christopher, 181n1

Frith, Uta, 181n1

Gallagher, Catherine, 184n30

Garrick, David, 62

Gaynor, Mitzi, 127

Gelman, Susan, 147

Gervais, Ricky, 103, 104, 106, 110

Gevorkian, Natalia, 189n9

Gibson, Mel, 38, 40, 42

Gimme Shelter (documentary), 109

Goffman, Erving, 15, 183n25

Goldman, Alvin I., 181n6

Gombrich, Ernst, 31, 186n16
Gomez, Juan, 181n1
Grant, Cary, 69
graphic narratives, 12, 157
Greenblatt, Stephen, 184n30
Greuze, Jean Baptiste, 158; *La Piété filiale*, 161–62, 164
Gross, James J., 186n22
Grossman, Tobias, 182n16
Guthrie, Stewart, 183n17
Guys and Dolls (film), 130
Gypsy (film), 133

Hammerstein, Oscar, II, 126
Hart, F. Elizabeth, xii, 188n1
Hearts and Minds (documentary), 109
Hegel, Georg Wilhelm Friedrich, 183n27
Hemingway, Ernest, *A Farewell to Arms*, 43, 187n32
Herman, David, xii, 188n1
Hickok, Gregory, 182n8
Hitchcock, Alfred, 153; *Notorious*, 69–72, 80–82
Hoffmann, E. T. A., 186n21
Hogan, Patrick Colm, xii, 188n1
Hogarth, William, 152
Holquist, Michael, xiii
Hornby, Nick: *Fever Pitch*, 72–75, 165; *How to Be Good*, 56–58
Hume, David, 182n9
Humphrey, Hubert, 107
Hunt, Helen, 40
Hutcheon, Linda, 191n3, 191n12
Hutcheon, Michael, 191n3, 191n12
Hyman, Steven, 184n28
Hynes, Catherine A., 181n3

Jackson, Tony, xii, 188n1
Joel, Billy, 143
Joe Schmo Show, The (TV series), 121
Jonas, Hans, 183n19
Jones, Wendy, 187n4

Kanwisher, Nancy, 181n1
Kaufman, Andy, 19, 113–14, 185n36
Kaufmann, Walter, 182n9
Keen, Suzanne, xii, 188n1
Keenan, Julian Paul, 181n1

Kelly, David J., 182n15
Kennedy, John F., 107
Keough, Peter, 190n10
Kern, Stephen, xiii, 169–72, 195nn1–3
Keysers, Christian, 181n7
Kleiner, Fred S., 182n14
"Knight Who Made Cunts Speak, The" (fabliau), 38–39, 40
Kramnick, Jonathan, xii, 188n1
Kuleshov, Lev, 80, 86
"Kuleshov effect," 80–81, 84
Kurzban, Robert, *Why Everyone (Else) is a Hypocrite*, 28–29, 32, 185n8, 185n10, 191n5

Lady Windermere's Fan (Lubitsch), 66–69, 72
Lane, Anthony, 86, 188n4
Lanser, Susan, 192n18
Leonardo da Vinci: *The Last Supper*, 149; *Mona Lisa*, xi, 19, 152, 154
Levin, David Michael, 183n19
Lindenberger, Herbert, 192n9
Lockerman, Gloria, 99–100
Longin, Sheryl, 121
Lopate, Phillip, 185n6
Lubitsch, Ernst, *Lady Windermere's Fan*, 66–69, 72
Luo, Yuyan, 182n11

Mamet, David, 83–84
Mansing, Howard, xii
Marshall, Rob, 137
Maysles, Albert, 109, 189–90n10
Maysles, David, 109, 189n10
Mazzetti, Mark, 189n7
McConachie, Bruce, xii, 188n1
McGinn, Colin, 79, 185n12, 186n28
McLane, Betsy A., 107, 189n4
McMillin, Scott, 125, 191nn2–4
Mead, Rebecca, 192n21
Merchant, Stephen, 104, 106
Merleau-Ponty, Maurice, 182n9
Merrily We Roll Along (musical), 134–35
Meyers, Seth, 92
Michaels, Walter Benn, 45
Midwood, William Henry, *At the Crofter's Wheel*, 170
Miller, Scott, 191n7

mind reading. *See* theory of mind
Mirren, Helen, 86
mirror neurons, 3–4, 192n9
montage, 80–81
Mozzhukhin, Ivan, 86
My Fair Lady (film), 126, 132–33, 142

Nettle, Daniel, 181n1
Newcomb, Anthony, 191n12
New Yorker, 92
New York Times, 92
No Lies (film), 109
Notorious (Hitchcock), 69–72, 80–82,
 85
Nussbaum, Martha, 182n9

Obama, Barack, 91–92
Ocean's Eleven (film), 90
Office, The (TV series), 103–6, 110–11,
 113, 117
Olivier, Laurence, 130–31
opera, 139, 140, 143

paintings, absorptive: and embodied
 transparency, 145–47, 158–61, 166–
 67; and patterns of mind reading,
 152, 154–55; and rules of transience
 and contrasts, 156–57; and sentimen-
 talism, 161–66
Palahniuk, Chuck, *Fight Club,* 49–53
Palmer, Alan, xii, 42–43, 184n31,
 185n4, 186n28, 188n1
Pemberton, Gayle, 99–100
Pentland, Alex, 183n24
Persson, Per, 79
Phantom of the Opera (musical), 126
Phelan, James, xiii, 187n7
Phelan, Peggy, 184n29
Picasso, Pablo, 149
Plantinga, Carl, 188n1
Plato, xi
Pleistocene, 2, 121, 122
poker faces, 90–92, 94, 97
Portillo, Isabel Jaén, xii
Practical Illustrations (Siddons), 62–63
Premack, David, 181n1
Priborkin, Klarina, 5–6, 182n10
Pride and Prejudice (Austen), 32, 33–36,
 37, 40, 85

Pride and Prejudice (film), 35, 36, 85
Primary (documentary), 107
Princess Diana, 87, 88, 89, 90
problem pictures, 173–76
proposal compositions, 169–72; and
 anxiety about female sexuality,
 173–76
Pugh, Ken, xiii

Queen, The (Frears), 86–90, 96
Queen for a Day, 119
Quigley, Rachel, 189n9
Quiz Show (film), 93–100

"Raggle Taggle Gypsy-o" (song), 142,
 192n19
Rains, Claude, 69
Ramachandran, V. S., 193n14
reality television, 117–22
Redford, Robert, 93
Reed, Lou, 143
Remnick, David, 92
restraint, rule of, 30, 33, 120, 180; in
 Bridget Jones, 38; in *Casablanca,* 83–
 85; in film, 82–86; in *My Fair Lady,*
 132; in *New York Times,* 92–93; in
 Notorious, 82, 85; in *The Office,* 104,
 105; and poker faces, 90–93; in *Pride
 and Prejudice,* 35–36, 85; in *The
 Queen,* 86–90; in *Quiz Show,* 95–98;
 in reality TV, 120; and sociocognitive
 complexity, 33, 85, 90; in theater, 84;
 in visual art, 155–57
Reza, Yasmina, *Art,* 151, 193nn12–13
Richardson, Alan, xii, 188n1
Richardson, Samuel, *Clarissa,* 56, 59–
 60, 61, 63, 76, 158
Rizzolatti, Giacomo, 181n5
Roach, Jean, 108
Roach, Joseph, 183n26
Rodgers, Richard, 126
Rolling Stones, 109
Rounders (film), 90
Rousseau, Jean-Jacques, *Emile,* 47–49,
 52
Rubens, Peter Paul, 152
Rubin, Philip, xiii
Ruppin, Eytan, 181n4
Rushes (film), 109